HELPING YOUR CHILD AT HOME

. . . with Mathematics

Macmillan/McGraw-Hill
School Publishing Company

EDITORIAL AND PRODUCTION STAFF

Editorial Director: Catherine White
Executive Editor: Ruth B. Cochrane
Editorial Production Manager: Helen Mischka
Editors: Barbara A. Everett, Lisa K. Higgins, Patricia S. Haupt

All photos on cover courtesy of Witherspoon Design and Will Shively Studio.

Macmillan/McGraw-Hill School Division
866 Third Avenue
New York, New York 10022

Printed in the United States of America

Level 5 ISBN 0-675-13097-2

9 8 7 6 5 4 3

Dedication

Macmillan/McGraw-Hill School Publishing Company is pleased to dedicate these materials to the following people, who have been our inspiration:

For America's classroom teachers, who strive for excellence in education and who need as much support at home from parents as they can get . . .

For America's parents and guardians, who seek better ways to support and interact with their children . . .

And principally for America's schoolchildren, who face a future filled with wondrous opportunity and serious challenge, and who need all of the educational resources we can provide as they carry our nation's hope and aspirations into the future.

Foreword

A MESSAGE TO PARENTS AND GUARDIANS

This may well be the single most important book you've ever used. By spending 10 minutes a day with your child doing the activities in this book—you can make a difference in your child's ability to learn more, learn faster, and get better grades.

There is indisputable evidence that children whose parents or guardians help them at home and stay in touch with their classroom teachers score higher than children whose parents are not involved.

- You don't have to have a high school diploma or college degree to help.
- It doesn't matter if you're rich, poor, or somewhere in between.
- It doesn't matter whether you are a white parent or guardian, a minority parent or guardian, or a single parent.
- It doesn't matter whether you live in a big city, a suburb, a small town, or a rural area.

IF YOU ARE WILLING TO COMMUNICATE WITH YOUR CHILD'S CLASSROOM TEACHER AND SPEND ONLY MINUTES A DAY WITH HOMEWORK AND THE ACTIVITIES IN THIS BOOKLET—YOU CAN HELP YOUR CHILD DO BETTER IN SCHOOL.

Your Child's Future Is at Stake

There are thousands and thousands of dedicated classroom teachers and educators across the United States who strive to provide the best possible education for your son or daughter.

But TEACHERS and YOUR CHILD NEED YOUR HELP to succeed.

Your child's classroom teacher devotes an average of six hours a day guiding your child to learn. Like you, teachers want your son or daughter to be able to get a good job and earn a decent living when they leave school. Teachers also want to make sure our children have the academic skills they need if they want to go on to college.

You Can Make a Difference in Your Child's Education . . . . And It Takes Only About 10 Minutes a Day!

Macmillan/McGraw-Hill's *HELPING YOUR CHILD AT HOME...with Mathematics* home study activities are designed to improve your child's understanding of math, permit you to participate in the activities with comfort and ease, and foster communication between you and your child.

- Parents do not have to have knowledge of math to do the activities.
- Each activity takes only about 10 minutes to do.
- Parents and their child do the activities together.
- Each activity includes easy-to-follow, step-by-step directions.
- You don't have to buy any materials to do the activities.
- You and your child will have fun doing the activities.
- Each activity reinforces a math concept your child is learning in school.

Every Child Needs Math Skills

Besides being able to read and write, math is arguably the most important skill your child uses every day of his or her life.

From the very basics—learning how to count, to tell time, to read a calendar . . . then to add numbers, multiply and divide—your child needs math skills to survive. Later on, your child will learn higher level math skills that will enable him or her to solve complex problems, to reason, and to acquire skills that will enable your son or daughter to get a good job.

During these early years, you can help prepare your child for the future. As a youngster, when your child sees that you value the knowledge of math, he or she will value it too.

We urge you to do the activities in this booklet with your child. Talk often to your child's classroom teacher. And make sure your children do their homework. The rewards will be great.

Acknowledgements

The Macmillan/McGraw-Hill School Publishing Company's *HELPING YOUR CHILD AT HOME...with Mathematics* was conceived by Chauncey A. Cochran and Stephen G. Clapp and developed in cooperation with Macmillan/McGraw-Hill and Cochran/Clapp, Inc.

Learning Design Associates, Inc., of Columbus, Ohio, provided editorial and production services for the development of *HELPING YOUR CHILD AT HOME...with Mathematics*. We gratefully acknowledge the help of John Buterbaugh, Project Manager, and Benjamin Shriver, Production Editor.

EXECUTIVE PROJECT EDITORS
Stephen G. Clapp
Chauncey A. Cochran

MANAGING AUTHOR
Richard W. Dykes, Austin, Texas

LEAD AUTHORS
Barbara Forth, Austin, Texas
Elizabeth Heath, Austin, Texas
Suzanne Kelley, Austin, Texas
Marilyn Neil, Birmingham, Alabama
Dwight Galbi, Lake Oswego, Oregon
Andrew Baumgartner, Augusta, Georgia
Lauren Caywood, Augusta, Georgia
Kay LeRoy, Augusta, Georgia
Jeanne Stroebel, Augusta, Georgia

CONTRIBUTING AUTHORS

TEACHER	SCHOOL	LOCATION
Faye L. Adams	Seven Hills School	Owensboro, Ky
Joy Allen	Sudderth Elementary	Monahans, TX
Barbara J. Arends	New Hope Elementary	New Hope, AL
Pennie Ascowitz	Pinewood Elementary	N. Lauderdale, FL
Jackie Barnes.	Eagle Heights Elementary	Azle, TX
Sarah M. Barranco.	Fultondale Elementary	Fultondale, AL
Marcelline B. Barry.	Jacksonville High	Jacksonville, FL
Patricia M. Basden.	Goliad Elementary	Odessa, TX
Linda E. Beal.	Taylor County High	Campbellsville, KY
Betty H. Benfield.	A.G. Cox Middle School	Winterville, NC
Margery J. Bessler.	Whittier Elementary	Cincinnati, OH
Bernice A. Best.	Beechgrove Elementary	Independence, KY
Diane H. Boesch.	Hurlbutt Elementary	Weston, CT
Ann Boisse.	Santa Rita Elementary	San Angelo, TX
Mary E. Bolin.	Lackawanna Elementary	Jacksonville, FL
Bruce A. Boling.	Eastwood Middle School	Indianapolis, IN
Diana Brettschneider.	Rockdale Elementary	Cincinnati, OH
Judy A. Brown.	Pleasant Valley Middle	Brodheadsville, PA
Jane Campbell.	Caney Elementary	Pippa Passes, KY
Carole H. Carroll.	Skyland Elementary	Tuscaloosa, AL
Mary Christensen.	Lamkin Elementary	Houston, TX
Jeannie Churchwell.	Norwood School	Birmingham, AL
Nicholas A. Cipparone.	Bardstown Middle School	Bardstown, KY
Robin-Lynn Clemmons.	Holy Innocents' Episcopal	Atlanta, GA
Carolyn Conklin.	Bess Race Elementary	Crowley, TX
Kay Cook.	Alief Independent School	Alief, TX
Deborah D. Coons.	Masonville Elementary	Utica, KY
Joyce Copeland.	Elizabethtown State Vo-Tech	Elizabethtown, KY
Carol Coulter.	Clearwater Elementary	Clearwater, KS
Maretta Countess.	Ed White Middle School	Huntsville, AL
Bonnye Cox	Keystone Heights Elementary	Keystone Heights, FL
Odessa Jackson Craig	Riverchase Middle	Birmingham, AL
Mary A. Davis.	Retired--Scarborough Middle	Mobile, AL
Regina P. Davis	Huntington Place Elementary	Northport, AL
Rebecca W. Dewitt	Eustis Middle School	Eustis, FL
Patricia C. Dickman	Oxford Junction Elementary	Oxford Junction, IA
Kathleen Eagleson	Trigg County Elementary	Cadiz, KY
Bonnie B. Ellerbee	Elem. Science Enrichment Ctr.	Griffin, GA
Melissa Elmer	Inter-city Baptist Elementary	Allen Park, MI
Marita H. Eng	Sandlewood Jr./Sr. High	Jacksonville, FL
George A. Erickson	Blessed Kateri Tekakwitha	Thoreau, NM

CONTRIBUTING AUTHORS

TEACHER	SCHOOL	LOCATION
Miriam M. Feinberg	Retired	Long Island City, NY
Dwayne Ferrell	Austin Peay State University	Clarksville, TN
Hope Friend	Wm. J. O'Shea Intermediate	New York, NY
Leigh Ann Gahn	Roberta B. Tully Elementary	Louisville, KY
Susan Goble	South Elementary	Brownwood, TX
Veronica Granger	Fulton County Elementary	Hickman, KY
Merdeces Guerra	T.M. Clark School	Portland, TX
Patsy Ann Gurganus	Reform Elementary	Reform, AL
Cina Guthrie	Woodrow Wilson Junior High	Dayton, TX
Karen E. Hampton	Wyoming Indian Elementary	Ethete, WY
Ann C. Haney	Southwest Elementary	Murray, KY
Mary D. Hankins	Highland School	Cedar Hill, TX
Jean Hanson	Andrews Elementary	Austin, TX
Carole Harris	Croton Elementary	Melbourne, FL
William R. Hastings	Retired	Berea, KY
Teresa Heath	Bethume Elementary	Haines City, FL
Terry Hefner	Sugar Loaf Elementary	Killeen, TX
Faye Hilgart	Blair-Taylor Middle	Taylor, WI
Patricia O. Hobson	Riverdale High	Riverdale, GA
Shirley A. Holbrook	Southern Elementary	Georgetown, KY
Michael Lane Hubbard	DeWitt Elementary	DeWitt, KY
Jennifer Huff	Glendale Elementary	Vero Beach, FL
Eileen Hughes	Pine Ridge Middle School	Naples, FL
Malanie Jennings	New Prairie Junior High	New Carlisle, IN
Arthur V. Johnson II	Nashua Senior High	Nashua, NH
Esther W. Johnson	Indian Trail Elementary	Louisville, KY
Sandra L. Johnston	Roland H. Lewis Elementary	Temple Terrace, FL
Janet Jolly	Erma Nash Elementary	Mansfield, TX
Eleanor R. Kane	Fishcreek Elementary	Stow, OH
Arnalynn Kellner	Stricklin Primary	Refugio, TX
Mary Rhonda Kendrick	Millard Elementary	Pikeville, KY
Jane Kenneth	Van Buren Junior High	Tampa, FL
Patrick D. Kent	Upper Arlington High	Upper Arlington, OH
Patricia K. Kessler	Quebec Heights Camas	Cincinnati, OH
Betty Koleilat	Meyer Elementary	Houston, TX
Rita S. Koontz	Radcliff Middle School	Radcliff, KY

CONTRIBUTING AUTHORS

TEACHER	SCHOOL	LOCATION
Beth Lazerick	Hebrew Day Institute	Rockville, MD
Jane Lechner	Evans Junior High	Ottumwa, IA
Rochelle Z. Lieb	Pinecrest Elementary	Immokalee, FL
Mary H. Lineberger	Woodbridge Elementary	Tampa, FL
Mary K. Longoria	Bruni Elementary	Laredo, TX
Jody Lowry	Lake Gibson Junior High	Lakeland, FL
Chicha P. Lynch	Capuchino High	San Bruno, CA
Marge Maginnis	Lincoln Ave. Elementary	Lakeland, FL
Jean M. Manzano	Brown Junior High	McAllen, TX
Hilda Martin-Wade	W.L. Smith Elementary	Petal, MS
Lois F. McMillan	Roger B. Chaffee Elementary	Huntsville, AL
Shelly H. McNanaro	Enterprise City Elementary	Enterprise, AL
Barbara J. Mead	Stevens Park	Dallas, TX
Faye Meads	Admiral Thomas H. Moorer	Eufaula, AL
Winnie Milan	Coronado Beach Elementary	New Smyrna Beach, FL
Diane Miles	Beulah High	Valley, AL
Becky Moore	Club Estates Elementary	Corpus Christi, TX
Ethel P. Morris	Skyland Elementary	Tuscaloosa, AL
Jean Morrow	Emporia State University	Emporia, KS
Peggy A. Neal	Rochambeau Middle School	Southbury, CT
Marilyn S. Neil	Retired--Jefferson City Elem.	Birmingham, AL
Sue Newman	Santa Anna Independent	Santa Anna, TX
Terri Orn	Bowie Elementary	Corsicana, TX
Robin Osborne	Greenville Elementary	Greenville, KY
Shirley Parker	Huntington Elementary	Huntington, TX
Beth Paslay	East Cliff Elementary	Portland, TX
Sarah W. Pass	Southeastern Elementary	Remlap, AL
Karen Pearce	Seabourn Elementary	Mesquite, TX
Kathie B. Peeples	Mountain Brook Junior High	Birmingham, AL
Florine Martin Perry	Retired--Mercedes ISD	Mercedes, TX
Michael Petrozzino	South Broward High	Hollywood, FL
Carol Rainbolt	Stephen F. Austin Elementary	Edna, TX
O.J. Robertson	Retired--Salem Elementary	Russell Springs, KY
Jan L. Robinson	Cooper Junior High	Buffalo Grove, IL
Vicki Rose	Crestwood Elementary	Crestwood, KY
William Rottmann	John Long Middle School	Grafton, WI

CONTRIBUTING AUTHORS

TEACHER	SCHOOL	LOCATION
Donna S. Salter	Evergreen City School	Evergreen, AL
Jill Salzer	Christa McAuliffe Elementary	Woodbridge, VA
Melanie Samuelson	Old Lawn/Robertson Elem.	Round Rock, TX
Lorna K. Brooks Sapp	Straub Elementary	Maysville, KY
Linda C. Schaftlein	Ruth Dunn Elementary	Louisville, KY
Bonnie Sexton	Gatesville Elementary	Gatesville, TX
Judith D. Slinker	Clayton Elementary	Clayton, AL
Harriette Smith	Palm Harbor Middle School	Palm Harbor, FL
Jonnie Smith	W.H. Burnett Elementary	Terrell, TX
Sarah C. Smith	Flour Bluff Elementary	Corpus Christi, TX
Connie Stallard	Evarts Elementary	Evarts, KY
Jimmy W. Stanley	Bankhead Middle School	Cordova, AL
Laura S. Stephens	Brewbaker Primary	Montgomery, AL
Pat Sullivan	Nob Hill Elementary	Ruidoso, NM
Doris Terry	Retired--Hatton Elementary	Town Creek, AL
Judith L. VanVoorhis	Muskingum College	New Concord, OH
Karen Veith	St. Columban Elementary	Loveland, OH
Susan Watson	Sparks Elementary	Pasadena, TX
Sherri Whisman	Powell County Middle School	Stanton, KY
Gerald White	Ohio State University	Columbus, OH
Brenda Whitsell	Summer Academy	Gallatin, TN
Louise E. Wilhite	Retired--Decatur City School	Moulton, AL
Carl B. Williams	Fl. Sch. for the Deaf and Blind	St. Augustine, FL
Deborah Williams	West Texas Elementary	Stinnett, TX
Pam Williams	Dixieland Elementary	Lakeland, FL
Bettye Wilson	Sally B. Elliott Elementary	Irving, TX
Karen Winn	Early Elementary	Early, TX
Judy Wolfsohn	J.W. Johnson Elementary	Hialeah, FL
Mary H. Wood	Boyd Buchanan School	Chattanooga, TN
Mary J. Wrenn	Saint Anthony School	Hawthorne, NJ
Ann Wyatt	Alexander Elementary	Duncanville, TX
Sara Yahr	Sally B. Elliott Elementary	Irving, TX
Cathleen M. Zucco	Le Moyne College	Syracuse, NY

GRADE 5

CONTENTS

SUGGESTIONS FOR USING THIS BOOK WITH YOUR CHILD

Here are a few simple suggestions for how you can use the ***HELPING YOUR CHILD AT HOME… with Mathematics*** series to help your child improve his or her mathematics abilities, perform better on tests, and earn better grades.

- If possible, set aside a specific time each day for you and your child to do an activity. You won't need a big block of time—most activities can be completed in 10 to 15 minutes—but it is a good idea to try to schedule the activity for a time when you are not likely to be interrupted.
- Work through the book at a pace that is comfortable for you. Start slowly, with one or two activities per week. Then, as you and your child become more familiar with the way these home-learning activities work, increase to three or four activities per week.
- Your child will benefit most if you work from the beginning of this book to the end. But don't worry about doing every single activity in the book. It's perfectly acceptable to skip one now and then if you don't have the necessary materials handy or if you think another activity looks more interesting. (And feel free to change or add to an activity based on your own personal experience!)
- When each activity is over, ask your child to tell you what he or she has learned or discovered. Talking in this way will help your child develop important thinking and communications skills.
- Encourage other family members to participate in the activities with you and your child. You might even consider making one evening each week a "family project night." Involving other family members will help develop your child's ability to work cooperatively in groups—and the shared sense of discovery will be a benefit for everyone in the family.

Part 1

Rounding and Estimating: using numbers to solve real-life problems

A note to parents

Have you noticed how big the numbers in the news are these days? Athletes sign contracts for $8,000,000. Books sell 250,000 copies in just weeks. Budget shortfalls are estimated in the hundreds of millions of dollars. These numbers may seem so large as to be unbelievable, but it's important for your child to be able to use them.

In this part your child will be using numbers as big as millions and as small as thousandths. To do this, your child will be practicing rounding and estimating.

Help your child handle the size of these numbers with some of these activities.

- Talk about how much you could buy with some really huge dollar amount, such as $500,000. For instance, how much peanut butter would $500,000 buy?
- Ask your child how many fifth-grade classes it would take to give each student one book from a total of 250,000 books. Discuss whether everyone in your whole town could each have a copy.
- Ask your child to estimate the total cost of items you are buying at the store, and then to estimate the amount of change you should get back.
- Explain to your child how you compare prices to find the best value for your money when you shop. Have your child help compare prices next time you both shop together.

Now practice rounding and estimating with the activities in Part 1.

Answer the Ads

Work with your parent to supply the numbers that answer these ads. You may want to refer to the place value chart for help. The place value chart helps you find the value of each digit.

PLACE VALUE CHART					
hundred-thousands	*ten-thousands*	*thousands*	*hundreds*	*tens*	*ones*
6	7	1	5	9	2

The value of each digit in 671, 592 can be explained as 600,000 + 70,000 + 1,000 + 500 + 90 + 2

Now answer these ads: (There may be more than one correct answer for some of the ads. You only need to find one.)

Wanted: A six-digit number using odd numbers only. Thousands place is greater than five.

____ ____ ____ ____ ____ ____

Wanted: The smallest number possible using all the digits 2 through 6 if 5 is in the hundreds place.

____ ____ ____ ____ ____

Wanted: A number where all digits 1 through 6 are used. Five takes the thousands place.

____ ____ ____ ____ ____ ____

Wanted: A five-digit number with three zeros. Ten-thousands place is four. Hundreds place is three.

____ ____ ____ ____ ____

BANK ON IT

To play this game, take turns with your parent picking 4 numbers from the bank. Write your numbers on your score card. Score a point for each rule your number matches.

- The number is less than five hundred thousand.
- The digit in the ten thousands place is greater than 5.
- The digit in the thousands place is even.
- The digit in the hundreds place is less than 4.
- The digit in the tens place is between 3 and 7.

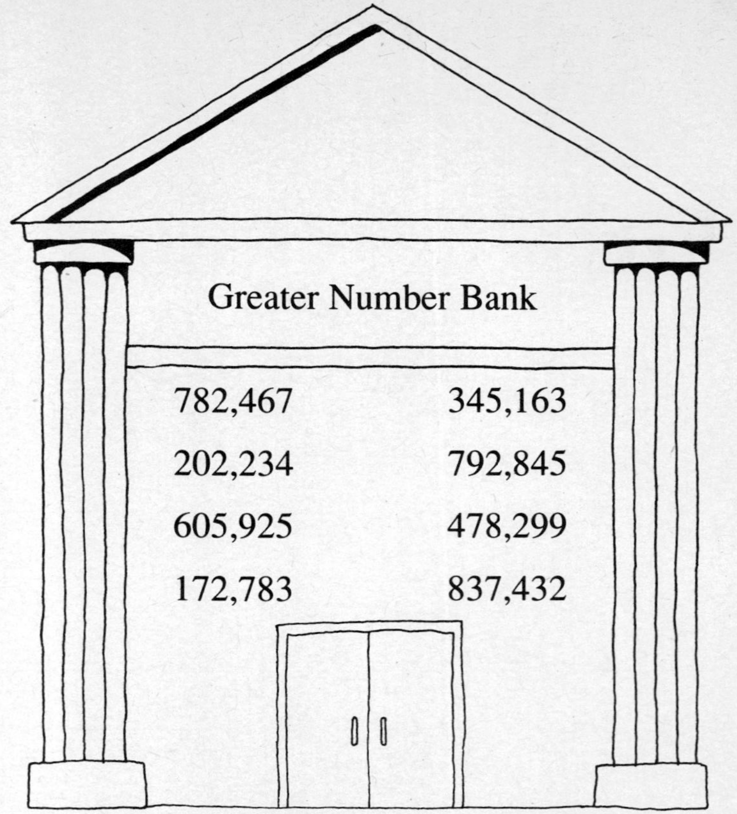

Name	
Number	*Points*
Total	

Name	
Number	*Points*
Total	

MILLIONS of Possibilities

Play this number game with your parent:

- Put the numbers in a bowl. Draw them out one at a time.
- Each of you takes a turn drawing the paper squares.
- Lay them out side-by-side in a line as you draw them, to make a 9-digit number. Start on the left and place each one to the right.
- Write your number on the game card. Don't look at the point counter until you've finished playing; then find your score with it.

You need: a set of cards or paper squares with the numbers 1 through 9, a bowl or cup

Points	*Point Counter (Don't read below, until game is over!)*
	If . . .
1	The number is less than nine hundred million.
2	The digit in the hundred thousands place is greater than 5.
3	The digit in the ones place is an even number.
4	The digit in the one millions place is less than 8.
5	The digit in the tens place is an odd number.
6	The digit in the ten millions place is 4.

Game Card	
My number	*Score*
___ ___ ___ , ___ ___ ___ , ___ ___ ___	
My parent's number	*Score*
___ ___ ___ , ___ ___ ___ , ___ ___ ___	

hundred millions, ten millions, one millions, hundred thousands, ten thousands, one thousands, hundreds, tens, ones

Numbers in the News

You need: newspaper, pen, sheet of paper, cup

Find ten 3- and 4-digit numbers in the newspaper and read them to your parent. Ask your parent to write them on strips of paper.

Fold the strips and put them in the cup.

- Draw a strip of paper out of the cup and read the number to your parent.
- Then tell how many thousands, hundreds, tens, and ones are in the number.
- Can you remember what the number stood for? (Example: 1428 votes)

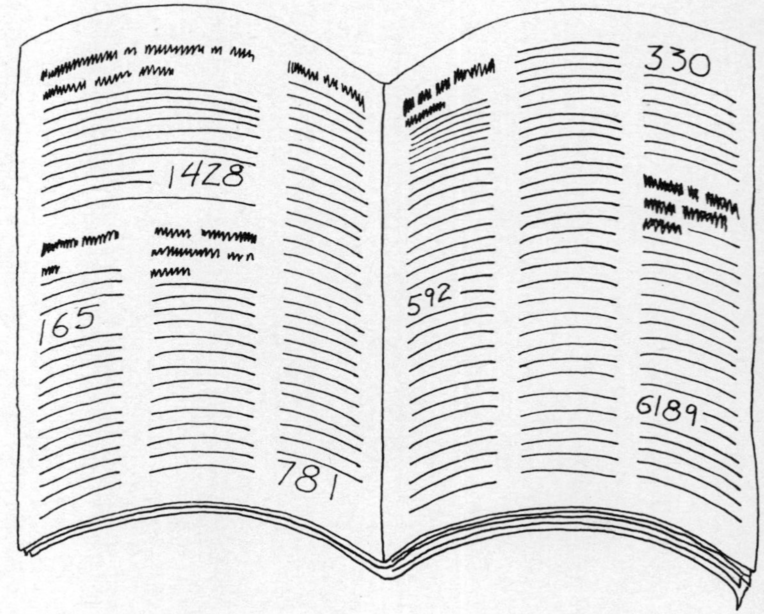

How many tens are there in 1428? How many ones?

How High the Moon!

There is a 40-foot flag pole in front of a school. Laid end to end, one hundred thirty-two flag poles like this one would stretch one mile.

Work with your parent to calculate how many flag poles laid end to end it would take to span the 3,273 miles from Seattle, Washington, to Miami, Florida. Use a calculator or pencil and paper.

How many poles, end to end, would be needed to reach the moon, which is 240,000 miles away from earth?

Name some items that you would be likely to see by the millions.

TOSS a ROUND

You need: number cube, paper and pencil

Practice rounding numbers with your parent this fun way.

Roll the number cube.

Each time the cube is rolled, write the number (in pencil) in any box on your game sheet.

When all the boxes are filled, round the numbers as directed.

Ask your parent to check your answers.

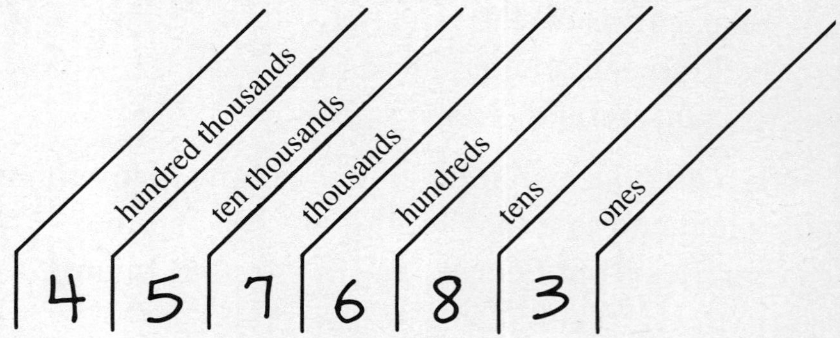

To the nearest thousand:
458,000

Game Sheet	
Number	*Rounded*
1. ☐ ☐ ☐ ☐	To nearest thousand
2. ☐ ☐ ☐ ☐ ☐	To nearest ten thousand
3. ☐ ☐ ☐ ☐ ☐ ☐	To nearest hundred thousand

For more practice, erase your game sheet and play again.

A Round Square Game

To play this game:

- Take turns with your parent.
- Pick any two numbers from the Number Pool. Round them to the nearest thousand and add the rounded numbers.
- Find your answer on the Game Square and put your initials in that box.

The first player with four in a row, column, or diagonal wins.

If you found a way to select the numbers that would give the answer box you wanted, describe it. If you didn't, see if you can figure one out.

Game Square

46,000	66,000	25,000	24,000
20,000	FREE	10,000	11,000
56,000	52,000	15,000	47,000
30,000	5,000	34,000	14,000

Number Pool

1,523	11,776
44,359	3,208
8,386	21,502

COLOR GRAPHICS

Ask your parent to help you gather data for a graph.

Make a tally of the colors of 25 items in or around your house. For example, you could count colors of socks, shirts, cups, or even the cars passing your house.

Keep your count on the tally sheet below.

Black		Green	
White		Red	
Brown		Yellow	
Blue		Gray	

Use your tally of colors to make a bar graph.

Make the bars horizontal, or left to right.

Show colors on one axis and numbers on the other. (Add to the numbers and the colors we began to list.)

Give your graph a name.

Which colors had the greatest number and which had the least number?

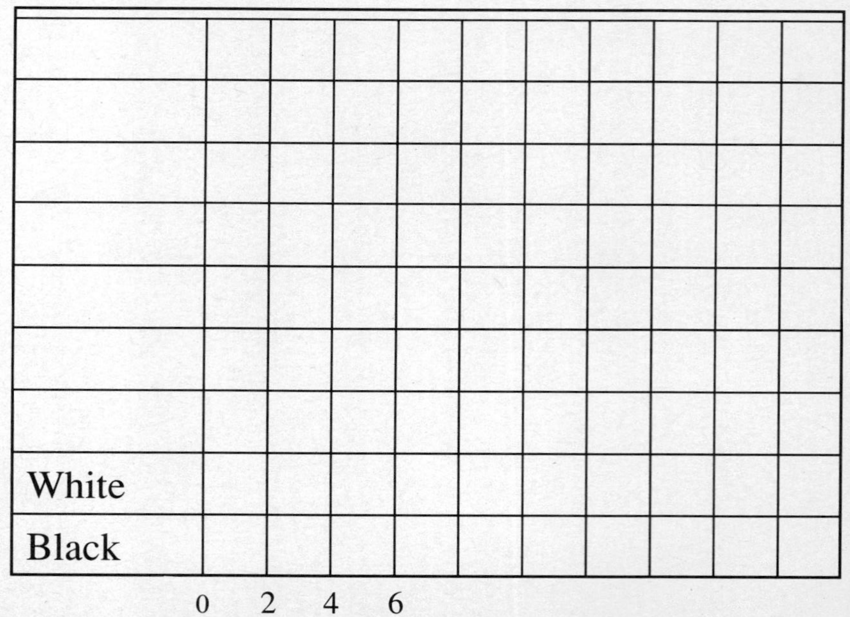

Sieving for Primes

Talk to your parent about prime numbers.

Remember that a prime number has no other factor except itself and one.

Both you and your parent estimate how many prime numbers there are up to 100.

Then here's how to sieve for primes, something like you can sieve for crayfish in a creek:

- Draw a ring around 2, and color the squares for all the multiples of 2 (4, 6, 8, 10, etc.)
- Ring the next uncolored number, and color the squares for each of its multiples that aren't already colored.
- Repeat the previous step until all numbers have a ring or color.

How many prime numbers did you ring?

Whose estimate came closer?

You need: a pen, a colored pencil

1	2	3	4	5	6	7	8	9	10
11	12	13	14	15	16	17	18	19	20
21	22	23	24	25	26	27	28	29	30
31	32	33	34	35	36	37	38	39	40
41	42	43	44	45	46	47	48	49	50
51	52	53	54	55	56	57	58	59	60
61	62	63	64	65	66	67	68	69	70
71	72	73	74	75	76	77	78	79	80
81	82	83	84	85	86	87	88	89	90
91	92	93	94	95	96	97	98	99	100

Why do you think this is called a sieve?

How Far to the Planets?

The planets of our solar system are divided into two groups. First come four small planets, the terrestrials: Mercury, Venus, Earth, and Mars. Outward, after a large space, come four giant-sized planets: Jupiter, Saturn, Uranus, and Neptune. (Tiny Pluto is in this region too.)

On a clear night, look for what appear to be very bright stars—they may be planets. Venus is the brightest planet. Mars, Jupiter, and Saturn also are often easy to see with the naked eye.

From books at home or in the library, learn how to recognize Venus, Mars, Jupiter, and Saturn in the night sky, and look for them with your parent. Also make a note of the distance each planet is from the Sun. Work with these distances on the next two pages.

GREAT DISTANCES (How Far to the Planets? page 2)

Enter your data into the chart below.

Mercury has been done for you.

Planet	*Distance from the Sun*
Mercury	36,000,000 miles
Venus	
Earth	
Mars	
Jupiter	
Saturn	
Uranus	
Neptune	
Pluto	

If the book you are working with gives the distances in kilometers, simply multiply each figure by 1.6 to convert to miles. Use a pocket calculator.

SEE HOW FAR (How Far to the Planets? page 3)

Name the planets orbiting the sun in the diagram below.

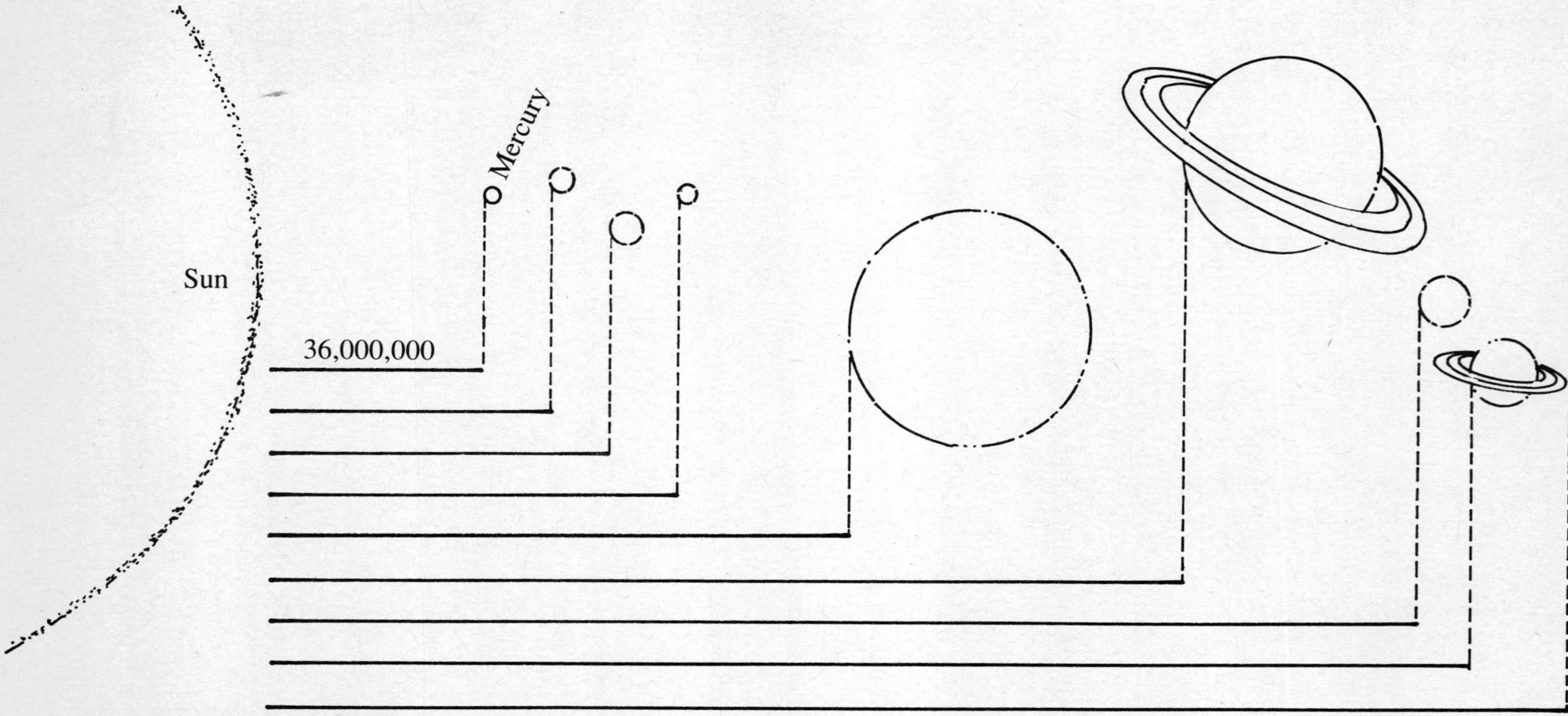

Ask your parent to help you write the average distance of each planet from the Sun. Mercury has been done for you (in miles).

YOU'RE RIGHT ON TARGET!

You need: 8 small items such as beans or grains of rice

Make a target, as shown, by drawing circles on a cut-open grocery sack. (You can use different size bowls to trace around.)

Lay the target on the floor.

Take turns playing with your parent like this:

- Each of you drop four beans, one at a time, on the target from a height of right below your nose.
- Mentally add your score and write your total down on a separate sheet.

At the end of ten turns, find the grand total for each person. High score wins.

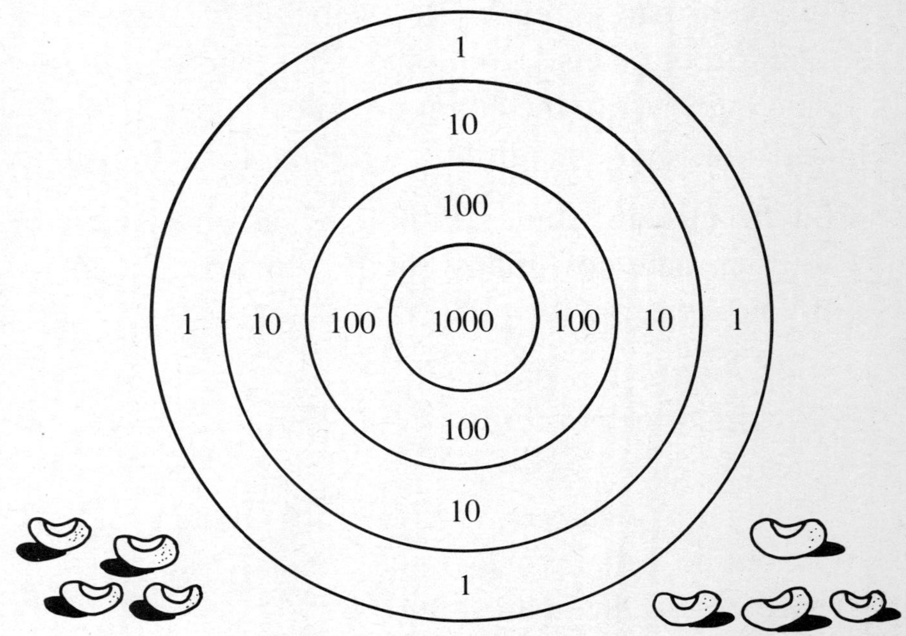

The Best Guess

You need: a package of dried beans or counters, a large bowl, 2 cups, a piece of paper

Place the beans or counters in a large bowl.

Have your parent reach into the bowl and take out a handful of beans. Count the beans and put them into a cup. Write the count down on paper. Estimate how many handfuls it will take to fill the cup.

You do the same thing. Put the beans into your cup. Estimate how many handfuls it will take to fill your cup.

Discuss with your parent:

- How many beans did your parent take, to the nearest ten?
- How many beans did you take, to the nearest ten?
- About how many beans did you two take together, to the nearest ten?

After rounding, add the actual amounts. How well did you estimate?

Repeat the activity.

About how many more or fewer beans did you pick up the second time?

If you were to draw four handfuls of beans, about how many would you pick up altogether?

PUT IT IN REVERSE!

Here's something fun to try with numbers.

A palindrome is a word that is spelled the same both frontwards and backwards. "Redivider" is the longest known one-word palindrome in the English language. Numbers can also be palindromes.

You and your parent find your own palindromes by following the directions below. After you read them carefully, race to see who can do it first.

- Each of you write a three- or four-digit number on your paper.
- Reverse the digits of the number and add the two numbers.
- If your answer is not a palindrome, reverse your total and add the two numbers. Continue reversing and adding until you find a number that reads the same frontwards and backwards!
- Try it again with new numbers.

Busy Numbers

Using the digits and letters on the dial of the telephone, race your parent to see who can find the following answers first. (Use paper and pencil when you have to.)

1. Add each column of numbers.
2. Add each row of numbers.
3. Add all the digits on the buttons.
4. Add the greatest 3-digit number you can make from the buttons to the least 3-digit number you can make.
5. Add T A L K to B U S Y.
6. Add S C H O O L to L E A R N.
7. Add your first name to your parent's first name.
8. Add the digit under the letters ABC, the digit below 6, the digit above the *, and the digit two down from the letters ABC.

Mental Minuses!

Shuffle the cards.

Divide the cards between you and your parent. Place the two groups of cards face down between you.

- Your parent turns a card. Then you turn a card.
- The one with the card with the larger number gives the difference between the two numbers. You may use pencil and paper.
- The difference is the score. Write it on the score sheet, in the proper column.

Continue until all cards have been turned. The player with the higher total score wins.

If you cannot give the difference, turn these two cards face down again.

Me	My Parent

You need: 20 cards with 2- and 3-digit numbers written on them

How do you subtract mentally?

PUT UP A SIGN

Here's a subtraction game to play.

Each player picks a sentence group below.

Put in subtraction signs to make true sentences in your group.

Check each other's answers.

Whoever has the most correct answers wins.

Group I

1. 1 0 9 7 4 5 8 = 639
2. 4 9 4 1 3 5 7 = 4,584
3. 7 4 1 8 1 2 5 = 7,293
4. 1 8 1 0 1 1 5 9 3 = 16,508
5. 6 5 2 0 5 2 4 7 8 3 = 40,422
6. 6 8 5 0 2 3 4 5 1 2 = 33,990

Group II

1. 1 6 9 6 8 9 8 = 798
2. 4 6 3 6 9 0 9 = 3,727
3. 4 7 7 2 1 0 5 9 = 3,713
4. 6 7 8 9 8 1 4 8 3 = 66,415
5. 4 1 8 5 1 9 5 1 = 40,900
6. 6 7 5 0 4 3 5 1 6 9 = 32,335

Tie-Breaker Problem: 9 8 7 6 5 4 3 2 1 0 = 55,555

(First to correctly place the sign wins!)

Rocket to Saturn

You need: 20 cards with subtraction problems you have made up on them (use up to 6-digit numbers), 2 markers for game pieces

Shuffle the cards and turn them face down.

Take turns playing with your parent like this.

- Draw a card. Find the difference. You may use paper and pencil.
- Use pencil and paper or a calculator to *check* the answer.
- Move one space higher each time you get a correct difference. If incorrect, do not advance.

The first player to reach Saturn wins.

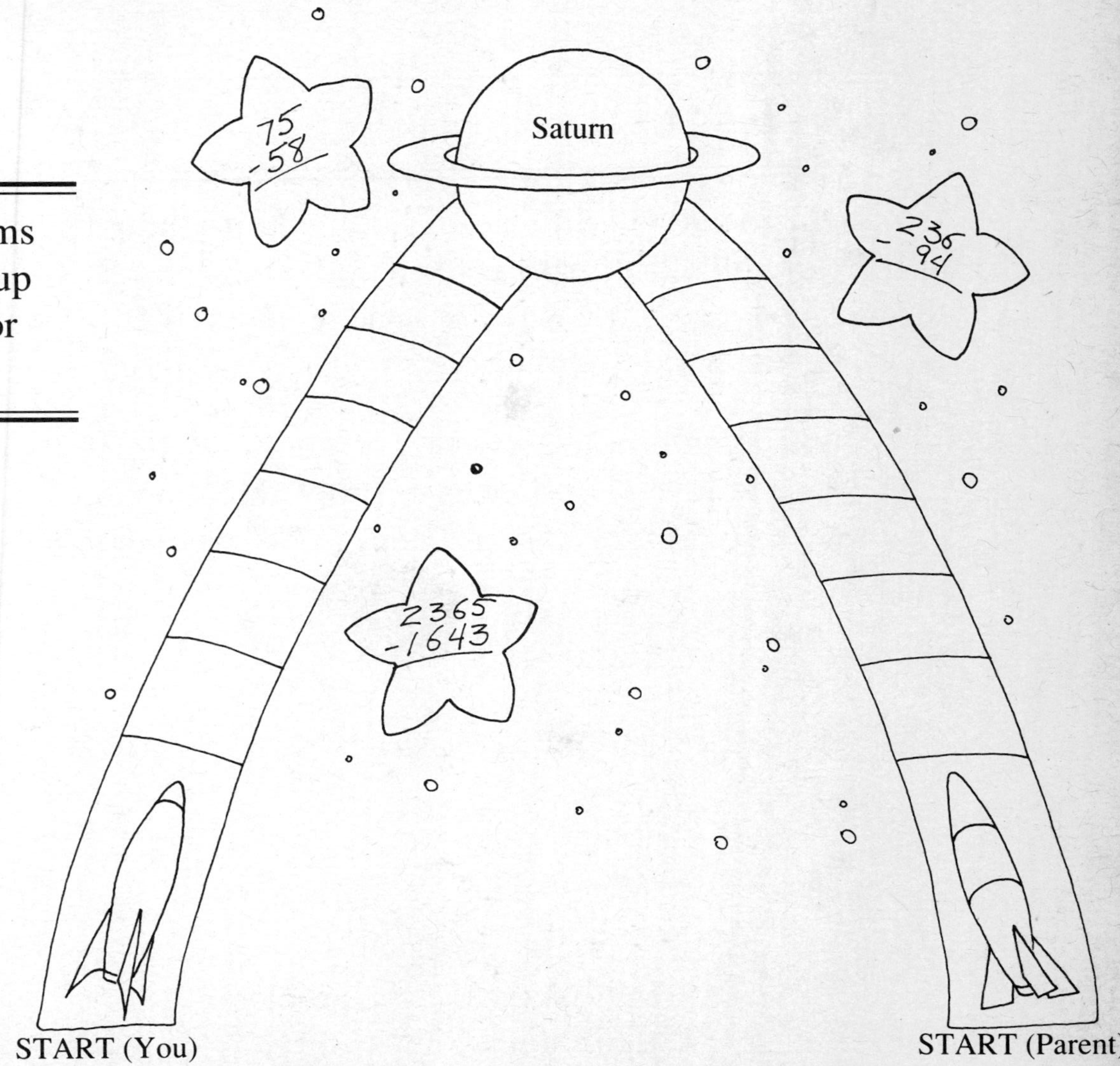

Million Dollar Question

You need: a $1 bill, an inch ruler

A mystery friend will give you $1,000,000 if you can say how far all the $100 bills that make up $1,000,000 would reach, placed end to end. (*Note:* a $100 bill is the same length as a $1 bill. Round to the nearest inch.)

First, think about how to solve this problem. Next, tell the steps you need to follow to work out this problem. Then, if your parent approves of your plan, write the steps to follow in this space.

Finally, follow the steps and work the problem.
Tell how many $100 bills, and win the million.

MIX AND MEASURE

Before you do this activity with your parent, look around your house for:

- Five things you would measure in centimeters
- Five things you would measure in meters

Write the name of each of the ten things on a slip of paper. Put them in a bag or bowl.

Now it's your parent's turn to do some metric thinking! Your parent draws out one slip at a time and tells how he or she would measure the object named on the paper, in centimeters or in meters.

Do you agree on the unit you should use to measure each object? Discuss your answers.

If you disagree on an object, talk about reasons for your difference and decide which unit is the better one.

m is the symbol for meter

cm is the symbol for centimeter

1 m = 100 cm

Together try to find five things in your house that could reasonably be measured in decimeters.

How Big Is That Room?

You and your parent measure the lengths of the walls in your room.
Use feet and inches or meters and centimeters. Find the perimeter.
Write it down. __________ $(p = l + l + w + w)$
Find the perimeter of another room in your house. Use the same units of measure. Record the perimeter. __________
Now find the answers to these questions by referring to your measurements. Tell your parent:

- Which room has the greater perimeter?
- How much larger is it?
- What would be the perimeter of a room the size of these two rooms combined? Hint: Think about the walls you will include.
- Make a drawing of a really great rec room that would be as large as these two rooms combined. Decide on a scale for your drawing, like 1 cm = 1m or $\frac{1}{2}$ in. = 1 ft. Draw in some things you would like to have in your rec room.

You need: a yardstick, meter stick, tape measure, or ruler

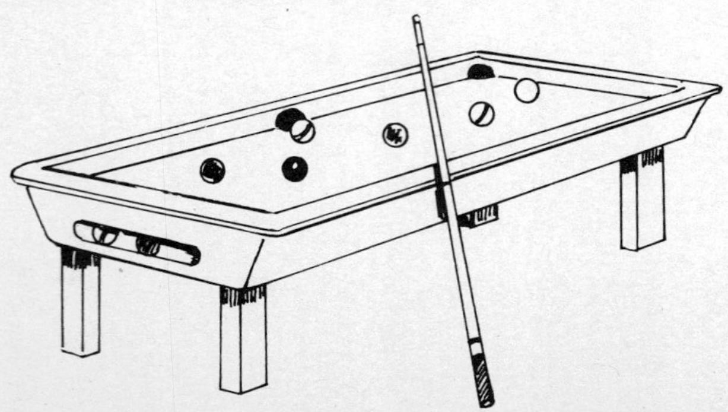

HOW FAR AROUND?

Look at each pair of shapes in box A with your parent. Estimate which of the two shapes has the longer outline (perimeter). Both you and your parent record your estimates. Do the same with box B.

Then measure the perimeter of each shape in centimeters and record it on the chart. (*Hint:* Each side of a square equals one centimeter.) The winner is the player who made more correct estimates.

A

Pair	*Shape*	*Perimeter Measurement*	*Shape*	*Perimeter Measurement*
A	1	☐ cm	2	☐ cm
B	3	☐ cm	4	☐ cm

B

What do people often build on the perimeter of their land?

Shop and Save!

You need: 2 index cards

The next time your parent is planning to make a small shopping trip to the grocery store to buy ten items or fewer, ask if you can make a list of the items needed. Make your list on an index card; label it like the one at right. When you go to shop, you'll be looking for savings on your items.

Item	*Original Price*	*Sale Price*
1. bread		
2. cheese		
3. milk		
4. eggs		
5. chicken		
6. fish		
7. plastic wrap		
8. potatoes		
9. bananas		
10. ice cream		
Total $		

KEEPING TRACK (Shop and Save, page 2)

While you are at the grocery store, look for savings! Enter the cost of each item in the appropriate space on your index card.

If there is an original price (before it was marked down), enter that amount also on the card, as in the example below.

Remember to keep the cash register receipt for later use.

Item	*Original Price*	*Sale Price*
1. bread	1.39	.99
2. cheese		2.45
3. milk	1.32	1.12

TOTAL SAVINGS (Shop and Save, page 3)

At home, calculate the total cost of your items.

(Compare the prices you wrote with the receipt.)

Enter the total on your card. Copy the labels on the top of the index card below, and calculate the amount of money you saved.

Item	*The difference between the original price and the price you paid*
1. bread	.40
2. cheese	—
3. milk	.20

How much would you have paid if all items had sold at the original price?

Ones, Tenths, Hundredths

Ones, Tenths, Hundredths

The object of this game is to try to build a bigger decimal than your parent. Here's how:

- Mix up the pieces of paper and lay them number side down on a table.
- Your parent picks up one piece of paper and places that number in any blank in the parent's score sheet below.
- Now it's your turn. Pick up a number and place it in one of your blanks.
- Continue until all blanks in the game are filled.

The winner is the one whose number has the greater value.

Play Game 2 and Game 3 and see who wins.

You need: 10 small pieces of paper (about the size of a postage stamp), labeled 0–9

	Me	*My Parent*
Game 1	___ ___ . ___ ___	___ ___ . ___ ___
Game 2	. ___ ___	. ___ ___
Game 3	___ . ___ ___	___ . ___ ___

Play these games again, but see who can build the <u>smaller</u> number this time.

BUILD YOUR NUMBER

You need: a number cube, four 3 × 5 cards

On each card, write one of the following: whole, tenths, hundredths, thousandths.

Lay the cards face down on the table and mix them up.

- Roll your number cube, then turn over a card.
- Write the *number* showing on the cube in the *position* written on the card. (For example, if you rolled a 4 and turned over the *thousandth* card, you would fill in one of your number blanks like this: _ . _ _ 4.)
- Turn the card face down again and mix it up with the others.
- Now it's your parent's turn to roll the number cube and turn over a card. Your parent should fill in his or her number blank and then mix up the cards again, just as you did.
- Take turns until all the number blanks are filled. When someone picks a card that cannot be used (because all those blanks are filled), then he or she gets to draw another card. Read the numbers aloud. The winner is the one who has the greater number.

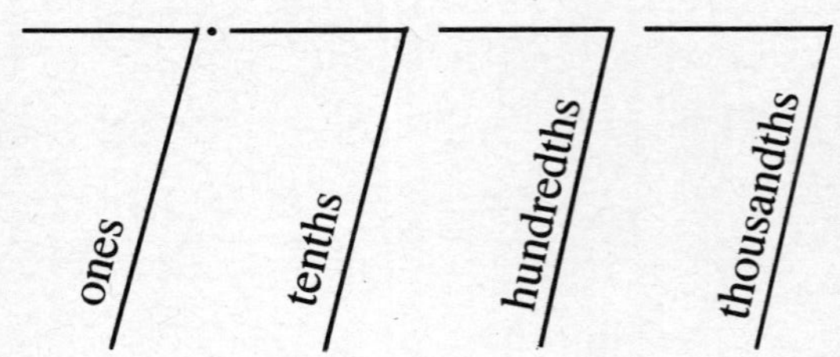

	Me	*My Parent*
NUMBER ONE	___.___ ___ ___	___.___ ___ ___
NUMBER TWO	___.___ ___ ___	___.___ ___ ___
NUMBER THREE	___.___ ___ ___	___.___ ___ ___

Is this a game of chance or skill?

TIC TAC DOUGH

You need: different markers (button, safety pin, eraser, penny)

In this game, your objective is to cover three numbers in a row that are *in order* from *greatest to least* or *least to greatest*.

Here's how to play:

- Your parent puts his or her marker on any square in Game One.
- Then you put your marker on another square in Game One. Remember to notice whether the numbers in a row (going across) or a column (going up and down) are *in order*. Notice the numbers in the diagonal rows, too.
- Keep taking turns until someone covers up three numbers in a row that are in order from least to greatest or greatest to least.
- Try again with Game Two!

$3.79	$3.65	$1.54
$.99	$1.00	$1.05
$.98	$1.05	$1.61

Game One

$0.01	$0.30	$0.25
$0.08	$0.10	$0.12
$0.09	$0.13	$0.11

Game Two

Jump Over

With your parent use the tape measure or yardstick to measure 100 inches on your driveway.

Use chalk to mark every 5 inches.

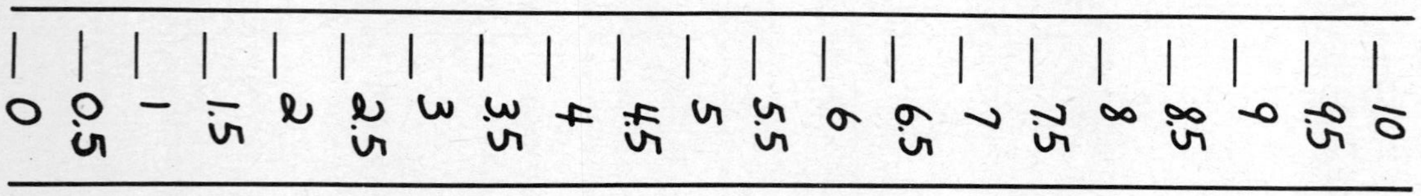

Label the first mark 0, the second mark 0.5, the third mark 1, the fourth mark 1.5, and so on until you reach 10.

Now line up behind 0 and JUMP! See how far you can go.

Measure to the back of your heel. Estimate your score to the nearest tenth.

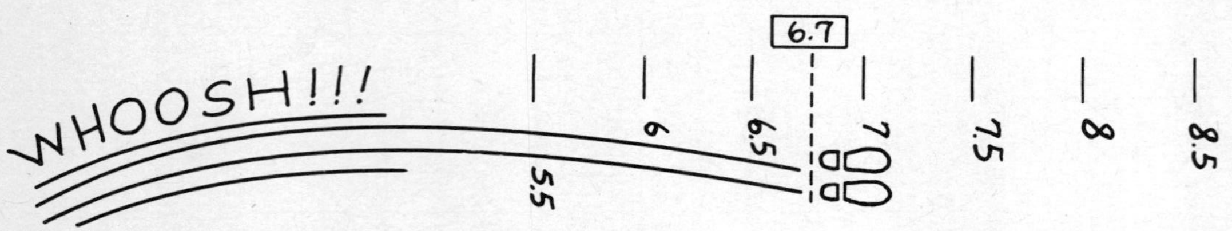

Now let your parent try. Measure that jump.

Take turns three more times.

Ask other family members or friends to jump. Make a chart and see who can jump the farthest!

You need: a sidewalk or driveway, a tape measure or yardstick, masking tape or chalk

How Many Kilometers to the End of the Rainbow?

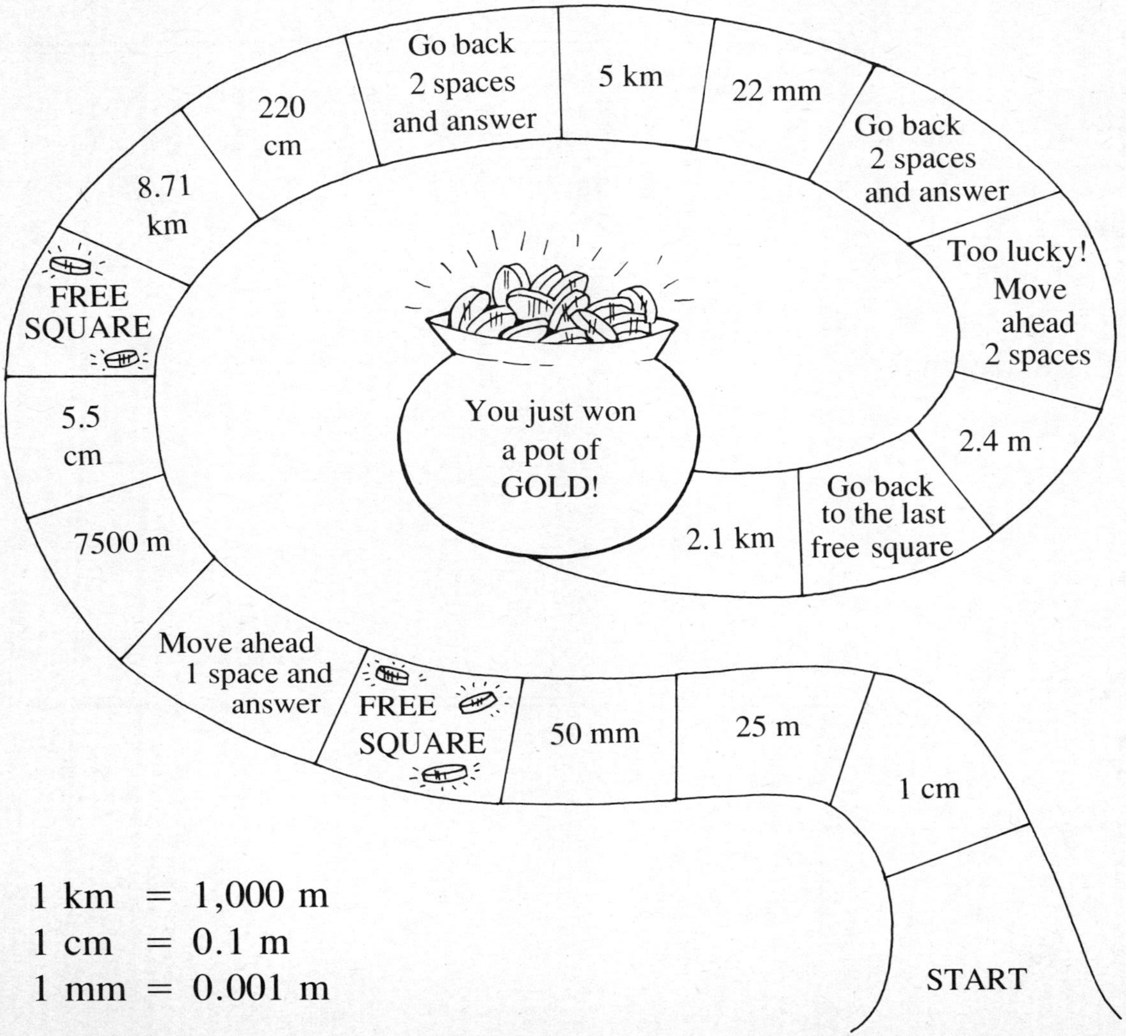

1 km = 1,000 m
1 cm = 0.1 m
1 mm = 0.001 m

You need: a number cube, 2 markers (buttons, beans, etc.)

How to play:

- Take turns with your parent rolling the cube.
- When you land on a space, you must rename the metric unit. For example: If you landed on a space that said 100 cm, you could answer 1 m or 1000 mm or 10 dm. *ALL WOULD BE CORRECT!*

If you miss an answer, you lose your next turn.

The first to get to the pot of gold wins, but the winner must do one especially *nice* thing for the other player!

Metric Concentration

mm cm m mm cm m mm cm m mm cm m

To play this game:

- You or your parent shuffles the cards. Lay them face down in rows.
- Take turns drawing two cards. If they are equivalent measurements, pick them up and turn over two more cards.
- The one with the most matches at the end wins.

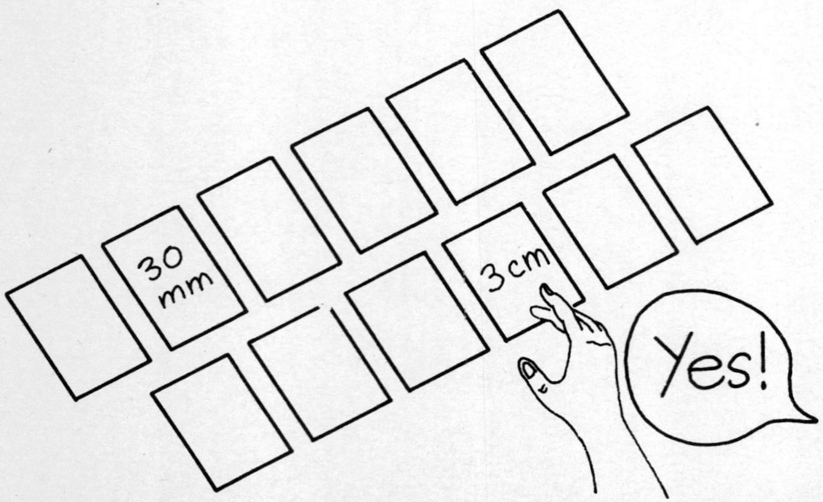

You need: 20 pieces of paper about the size of index cards. Label each card with one of the following:

23 mm	2.3 cm
3 cm	30 mm
40 mm	4 cm
2.5 cm	25 mm
5 cm	0.05 m
12.6 cm	126 mm
49 mm	4.9 cm
35 mm	3.5 cm
30 mm	3 cm
1.4 cm	14 mm

Can you change each card to meters?

$hopping $trategies

- Find advertisements for five things you would like to buy—like toys, clothes, or food. Write their names in the left column below. Write the *exact* cost of each item in the middle column below.
- Now *guess* how much those figures add up to and write in your guess at the bottom of the left-hand column.
- Next, round the cost of each item to the nearest dollar and write it in the right-hand column.

You need: a newspaper or brochure that came in the mail

- Add the numbers in the right-hand column. This is your *estimated cost.*
- Ask your parent to add the numbers in the middle column. This is the *actual cost.*
- Compare your guess, your estimated cost, and the actual cost. How close is your *guess* to your *estimate*? Which one is higher—your estimate or actual cost?

Item	*Exact*	*Rounded*
$ ______	$ ______	$ ______
$ ______	$ ______	$ ______
$ ______	$ ______	$ ______
$ ______	$ ______	$ ______
$ ______	$ ______	$ ______
Guess $ ______	Actual Cost $ ______	Estimated Cost $ ______

How many twenty-dollar bills would you need in order to pay for these five items? Would the cashier give you change? Estimate how much.

HEAD IN THE CLOUDS!

Take turns playing this high-flying game with your parent.

Here's how to play:

- Choose one decimal number from cloud 1 to add to a decimal number from cloud 2.
- Estimate the sum and write it on a piece of paper. Put your initials on that slip, too. Put the paper in the correct cup.
- Cross out the numbers in the clouds as you use them.
- Work with your parent to check each other's answers.
- Add your points to see how many you got. (Only count the ones that got in the correct cup.)

You need: 5 paper cups or other containers with these labels:

9 small pieces of paper, a pencil

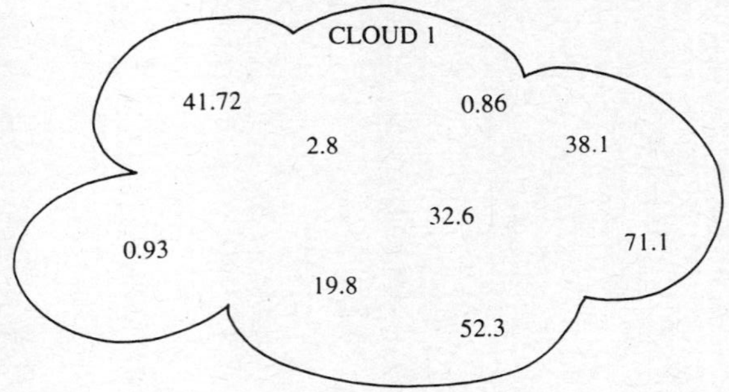

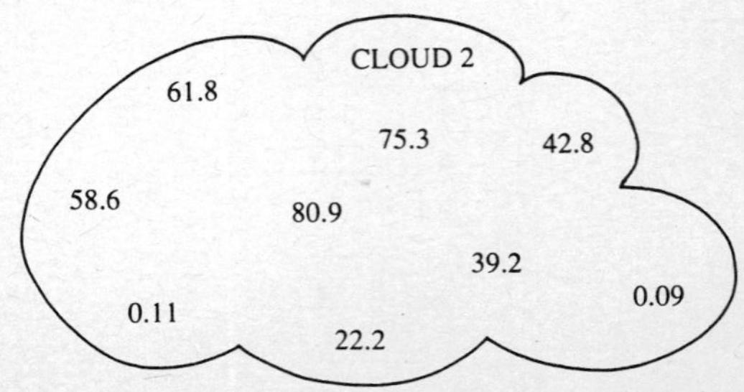

A DECIMAL FOR THE TEACHER

Take turns with your parent connecting the dots.

- *First,* connect dots next to each other with decimals that can be added together to equal 1.0.
- *Next*, connect dots next to one another with decimals that can be added together to equal 0.75.
- Can you solve the picture puzzle? If you can, connect the rest of the dots to complete the picture.

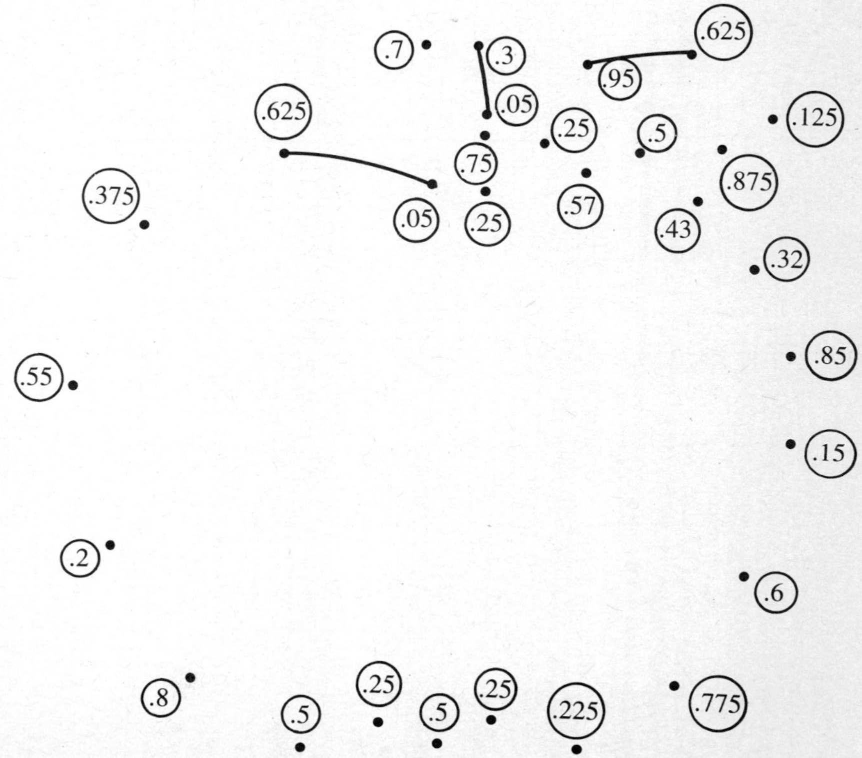

THE GREAT DECIMAL RACE

Make 18 small cards like these:

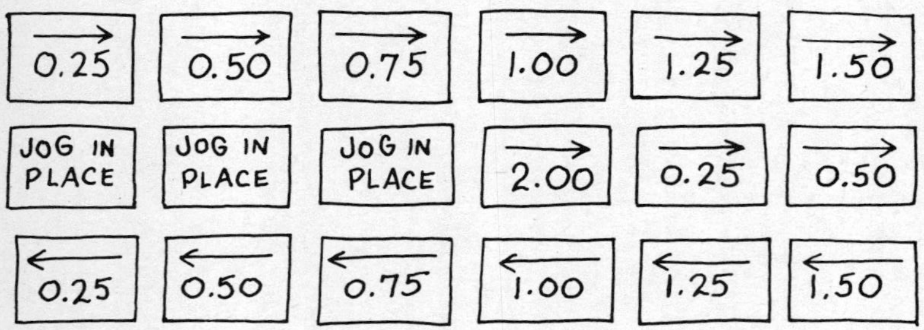

Shuffle the cards and place them face down in a pile.

Put your markers on 2.00.

You need: 18 small cards or pieces of paper, 2 game markers (button, paper clip, etc.), a pencil

Then, to race: Draw a card and move your marker in the direction of the arrow. Move your marker as far as the card shows. (If the card tries to send you all the way off the end of your line, don't go—just stay at the end.)

Whoever is closer to 4 when all cards are drawn, wins.

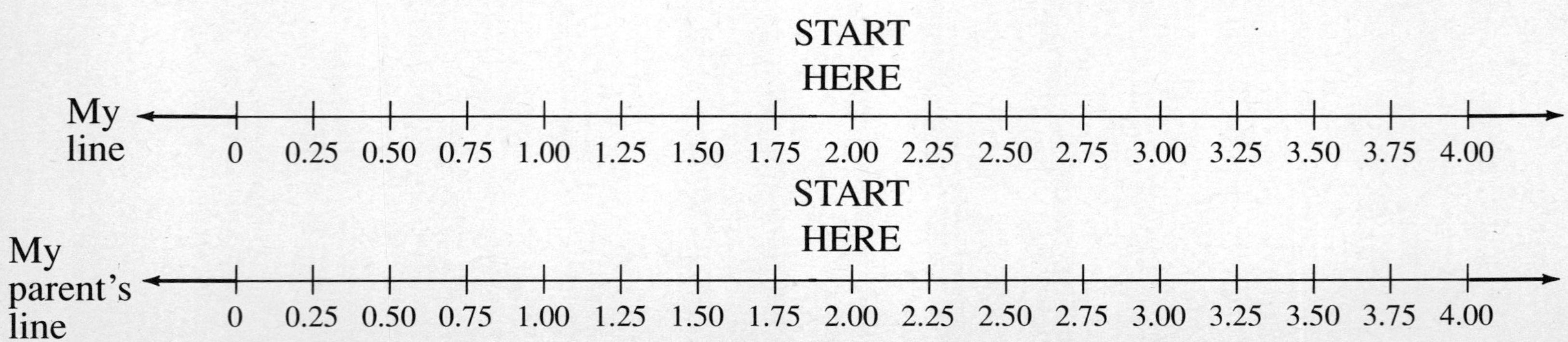

FRACTURED Bingo

You need: a bowl, small pieces of paper with the decimals printed on them that are shown at right, about 25 small markers like beans

My Card

$\frac{1}{2}$	$\frac{1}{10}$	$\frac{2}{5}$	$\frac{3}{4}$
$\frac{3}{10}$	$\frac{2}{4}$	$\frac{1}{4}$	$\frac{2}{10}$
$\frac{2}{2}$	$\frac{1}{5}$	$\frac{0}{2}$	$\frac{3}{5}$
$\frac{5}{5}$	$\frac{4}{10}$	$\frac{5}{10}$	$\frac{7}{10}$

My Parent's Card

$\frac{4}{4}$	$\frac{1}{5}$	$\frac{1}{10}$	$\frac{1}{2}$
$\frac{2}{10}$	$\frac{3}{4}$	$\frac{2}{2}$	$\frac{4}{5}$
$\frac{2}{4}$	$\frac{5}{5}$	$\frac{2}{5}$	$\frac{4}{10}$
$\frac{3}{5}$	$\frac{0}{2}$	$\frac{1}{4}$	$\frac{3}{10}$

0.5	0.25
0.5	0.75
0.5	1.0
0.5	1.0
0.1	1.0
0.2	0.0
0.3	0.7
0.4	0.8
0.6	0.9

To play:

Put the pieces of paper in the bowl. Mix them up.

- Take turns drawing out a decimal.
- Each time you match one of the decimals with an equivalent fraction on your card, put a marker on the fraction.
- The first to get 4 markers in a row (↕, ↔, or ⤢) calls out "Fractured Bingo!" That's the winner.

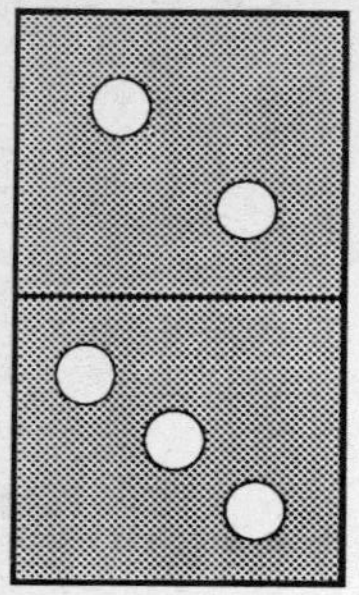

Decimal Dominos

You need: 20 small pieces of paper (about 1″ × 2″)

Make your decimal dominos by writing two numbers on opposite ends of each piece of paper.

Here are the numbers to write:

0.3	0.4	0.6	0.15	0.33	0.134	0.11	0.375	0.20	0.75
$\frac{6}{10}$	$\frac{1}{20}$	$\frac{1}{4}$	$\frac{2}{3}$	$\frac{1}{9}$	$\frac{3}{4}$	$\frac{7}{8}$	$\frac{1}{12}$	$\frac{2}{5}$	$\frac{1}{6}$
0.05	0.25	0.1	0.125	0.5	0.8	0.66	0.17	0.875	0.8
$\frac{4}{8}$	$\frac{1}{8}$	$\frac{3}{10}$	$\frac{1}{3}$	$\frac{3}{8}$	$\frac{1}{5}$	$\frac{4}{5}$	$\frac{2}{15}$	$\frac{3}{20}$	$\frac{1}{10}$

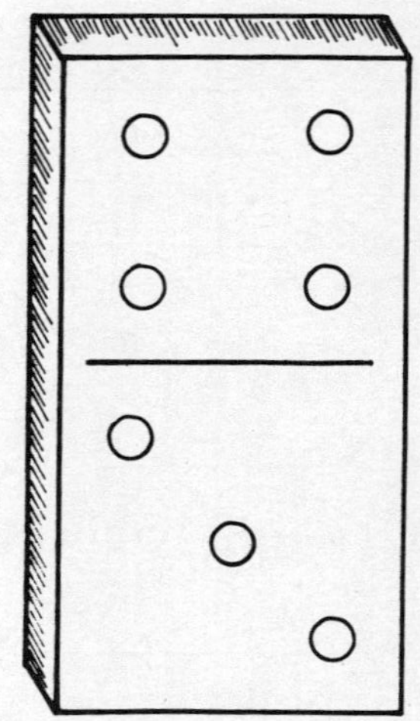

Turn your decimal dominos face down on the table and mix them up.

You and your parent pick 5 dominos each. Keep the numbers hidden!

Then, to play:

- Lay one of your dominos face up. Your parent has to match either the fraction or the decimal with an equivalent number. Then it's your turn. (You can play on either end.)
- If you can't make a match, you must pick another domino from the table.
- The first person to get rid of all his or her dominos is the winner!

If you like, mix the dominos up again and play another game.

Extended Activity

DECIMAL DETECTIVE

Go to the library with your parent to track down these decimal clues:

Look at a chart to find out, or ask the librarian, what *general subjects* are catalogued under these whole Dewey Decimal numbers: 500, 600, 700, 800, and 900.

Find out where children's and adults' books are.

Go to the shelves and find any one book that is listed under *each* general subject.

On the chart on the next page, write down the title and Dewey Decimal number of each book. You will have information on *five* books.

Read the complete instructions on the next page before you begin your search. (That way you won't have to go back and forth too much.)

SUBJECTS CHART (Decimal Detective, page 2)

When you find each book on the shelf, notice how all the books near it are on a similar *specific subject* (like computers or American History). Can you tell where the collection of books on this specific subject begins and ends? What *specific subjects* come before and after your book's specific subject on the shelves? In the right column write down the name and Dewey Decimal number of the first book you come to on the shelves that is about a different *specific subject* than the book you started with.
Would you like to check out any of these books? If you don't have a library card, maybe you can get one now. Ask the librarian.
When you get back home, complete the exercise on the next page.

General Subject	*Name and Dewey Decimal Number of Book 1*	*Name and Dewey Decimal Number of Book 2*
500		
600		
700		
800		
900		

PARAGRAPH AND PICTURE (Decimal Detective, page 3)

Write a paragraph to tell about what you learned about using decimal numbers for organizing books in the library. Do you think it is a good way to organize books? How are Dewey Decimals different from decimals that have mathematical value?

Then draw a picture that shows the subject of one of the books you found in the library. Did you check this book out to read? If so, you will be able to draw a very specific picture. Show it to your parent and explain what your picture is about.

My Paragraph	*My Picture*

Part 2

Multiplying and Dividing Whole Numbers: making sure there's enough

A note to parents

The next time there isn't enough pizza for everyone to have seconds, your child will suddenly become very much interested in multiplication and division. Why? It's because multiplication and division help people make sure everyone gets a fair share. These are two of the most practical math skills your child will ever learn, and your child will be practicing them in this section. Also in this section are some tricks for using mental math to multiply and divide some numbers without using pencil and paper.

Ask your child to help you make some simple, everyday calculations such as these:

- You need 3 potatoes for each family member for this week's suppers. How many potatoes should you buy? (**Multiply** the number of family members by 3.)
- You have $15 to spend for lunch at work this week. How much is that per day? (**Divide** $15 by the number of days in your work week.)

As your child becomes more comfortable with multiplying and dividing, encourage him or her to have a little more fun with creative activities such as these:

- Play mental math quiz games together. Use numbers that end in 0 or 5, and choose numbers that divide evenly with no remainders. Ask, "How many 25's are there in 75?" When your child is ready for harder questions, ask, "How many 25's are there in 7,500?"
- Ask your child to invent make-believe multiplication and division problems. For example, if 12 dragons each rule 12 magic cities, how many cities are in the whole magic kingdom?

With practice, your child's math skills will multiply before your very eyes!

Now try the activities in Part 2.

8 × Silly

This activity will give you some multiplication practice and a good laugh, too.

Take turns making word problems.

Choose one part from Box I and one from Box II. Some combinations will be pretty ridiculous! Write the problem.

Calculate the answer mentally. Then write the answer. Check each other's answers with a calculator or paper and pencil.

I

a. 42 students
b. 63 elephants
c. 28 nesting birds
d. 22 county road workers
e. a group of 19 dolphins
f. 16 city electricians
g. 52 acorns
h. 34 bakers

II

a. each laid 3 eggs
b. ate 6 pounds of squid apiece
c. each prepared 8 cherry pies
d. fell from each of 9 trees
e. each repaired wiring in 3 buildings weekly
f. sold 5 tickets each
g. each ate 4 bales of hay daily
h. each put up 7 traffic signs

The Match Game

You need: 10 colored pencils or crayons

Remember: Solve the problem inside the parentheses first.

Set 1: 36 | 45 | 48 | 32 | 15 | 56 | 63 | 18 | 24 | 12

Set 2:

$3 \times (7 + 5)$	$5 \times (6 + 3)$	$3 \times (10 + 6)$	$4 \times (4 + 4)$	$5 \times (1 + 2)$
$7 \times (4 + 4)$	$7 \times (6 + 3)$	$2 \times (5 + 4)$	$9 \times (5 + 2)$	$2 \times (10 + 8)$
$4 \times (4 + 2)$	$4 \times (10 + 2)$	$2 \times (10 + 6)$	$3 \times (3 + 2)$	$4 \times (10 + 4)$
$3 \times (20 + 1)$	$2 \times (10 + 2)$	$3 \times (4 + 2)$	$4 \times (6 + 3)$	$3 \times (10 + 5)$
$6 \times (6 + 2)$	$8 \times (2 + 2)$	$3 \times (4 + 1)$	$7 \times (6 + 2)$	$6 \times (2 + 1)$
$6 \times (2 + 2)$	$3 \times (2 + 2)$	$2 \times (3 + 3)$	$9 \times (2 + 3)$	$6 \times (1 + 1)$

- Take turns.
- Pick a number from Set 1. Choose a color and shade the box.
- Then find three expressions from Set 2 that show different ways to write the number. Shade the three boxes the same color.
- Your parent repeats the procedure using a different color.
- Continue selecting cards from Set 1 and the equivalent expressions in Set 2 until all boxes are shaded.

FACTO

You need: 23 cards to make multiplication cards, marker, beans for game pieces

Make a set of multiplication cards as shown below.

10 × 20	40 × 8	20 × 40	70 × 300
20 × 400	4 × 250	30 × 16	60 × 70
6 × 60	9 × 3000	7 × 200	8 × 90
80 × 80	30 × 60	50 × 90	60 × 90
24 × 300	50 × 50	200 × 50	800 × 2
3 × 150	9 × 700	5 × 8000	

F	A	C	T	O
480	5400	360	200	800
6300	1400	Free	6400	7200
21,000	1800	1000	4200	40,000

F	A	C	T	O
21,000	4500	7200	40,000	450
480	320	Free	1600	200
27,000	8000	2500	10,000	720

Each of you choose a Facto card. Shuffle the multiplication cards and place them face down. Turn them over one at a time. If the product is on your Facto card, cover it with a game piece (bean). Say, "Facto!" when you cover a row horizontally, vertically, or diagonally.

Explain how you use basic facts to multiply by 10, 100, and 1000.

MULTIPLES CROSS NUMBER PUZZLE

Complete the puzzle. Write the products in the squares.

Across:

A. 60×20
B. 20×40
C. 90×60
E. 4×1500
G. 60×500
H. 5×50
J. 10×400
K. 2×420
M. 33×100

Down:

A. 30×500
B. 80×100
C. 7×80
D. 4×500
F. 5×300
H. 40×60
K. 20×40
L. 2×220

Make your own cross number puzzle for your parent to complete.

Take Your Place in Line

Take turns with your parent.

Multiply to find each product. Then place a point on the number line for the product. Tell how you decided where to place the point. (The first one is done for you.)

A. 243
× 8
1944

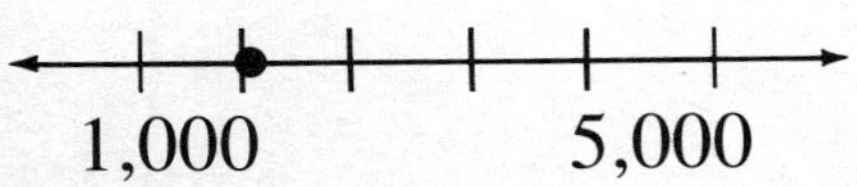

B. 316
× 12

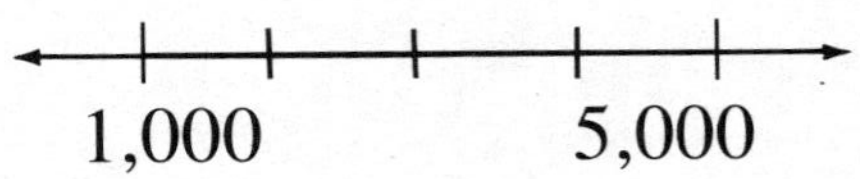

C. 972
× 9

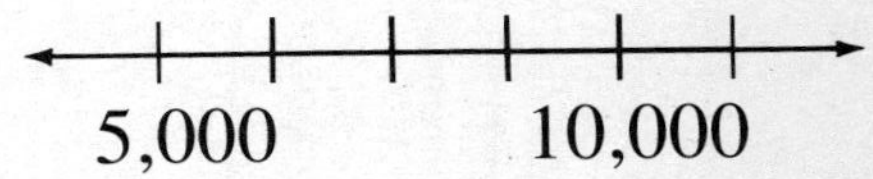

D. 2,081
× 5

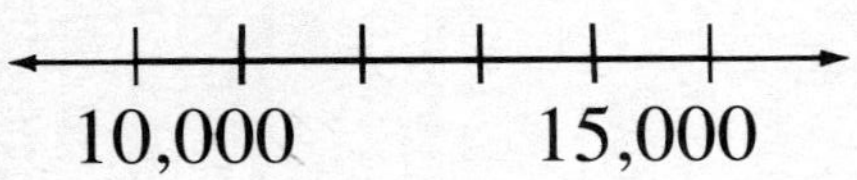

E. 4,519
× 3

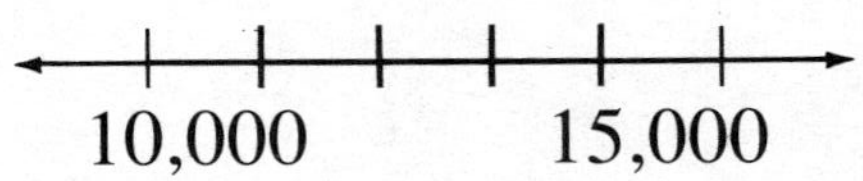

F. 3,703
× 16

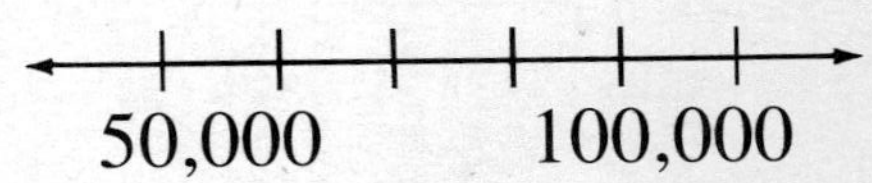

Draw a number line from 1,000 to 100,000 and place all the products on it. Use the letter of each problem to show that point on the line.

Multiplication Target

Take turns.

Pick one number from Set I and one from Set II.

Multiply to find the product.

Find that number on the target and put your *X* or *O* on it.

To win you must get a line of *X*s or *O*s going straight through the FREE space to the other side (5 spaces in a row).

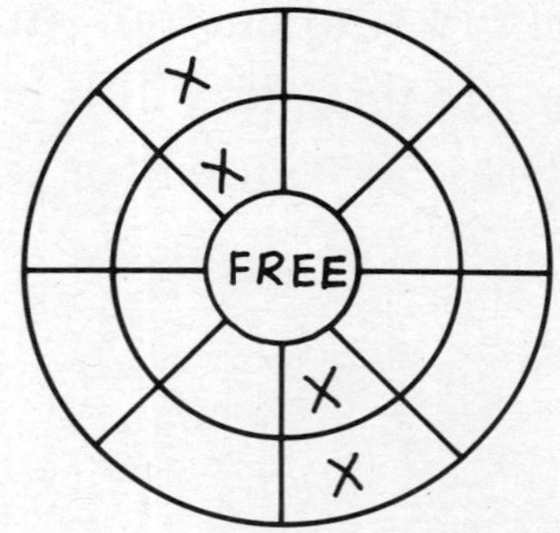

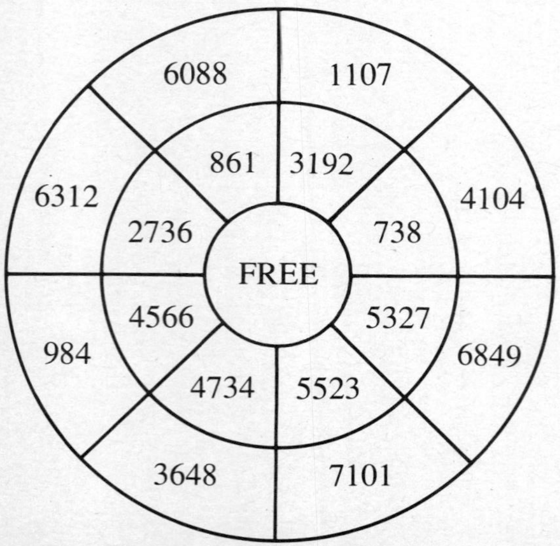

Set I
9
8
7
6

Set II
123
456
761
789

Toss a Factor

You need: 20 index cards or pieces of paper, a coin

Make 10 cards with 2-digit numbers. Write a 2 on the back of each.

Make 10 cards with 3-digit numbers. Write a 3 on the back of each.

Shuffle each deck and place them face down in 2 stacks.

Take turns.

- Toss the coin twice. Take a card from the stacks for each toss. ''Heads'' calls for a 2-digit number and "tails" calls for a 3-digit number.
- Both you and your parent find the product of the number cards.
- The one who gets the correct answer first gets 5 points. The other gets 3 points for a correct answer.

Play for 5 minutes. The higher score wins.

MULTIPLICATION MAZE

You need: 2 sets of cards with the numbers 1 through 9, small objects such as buttons or beans for game markers

Mix the cards and stack them face down.

Take turns. (As the cards are turned, make a discard stack to use when the other stack is used up.)

- On your first turn, put your marker on 372.
- Draw 2 cards to make a 2-digit number.
- Multiply this number by 372.
- If you are correct, move ahead 3 spaces. If you are incorrect, move back 1 space.
- Now your parent takes a turn.
- On your next turn draw 3 cards from the stack to make a 3-digit number. (The next time 2, then 3, and so on.)

The first player to the FINISH wins.

252	710	912	489	301
633	503	475	977	645
584	869	FINISH	380	126
151	386	541	772	601
203	816	459	372	START

LOOP THE PRODUCTS

You need: Use any two colors of markers or crayons

You and your parent take turns looping factors and their product. The factors and product may be vertical, horizontal, or diagonal.

You loop with one color. Your parent loops with the other color.

When all factors and products have been found, count the same color loops to find the winner.

8	3	24	2	5	8	4
5	2	10	3	2	6	8
40	6	16	7	24	2	32
3	5	15	5	2	12	4
9	7	5	35	8	14	6
5	25	21	25	16	40	24
6	2	12	4	5	3	15
30	10	4	3	8	5	40

A WEIGHTY Problem

You need: bathroom scales (If you don't have them, you can estimate weights.)

What to do first:

- Weigh yourself.
- Multiply your weight by 20.
- Is the product more or less than one ton?

What to do next:

- Weigh two other people in your family.
- Multiply their weights by 20.
- Are the products more or less than one ton?

1 ton = 2,000 pounds

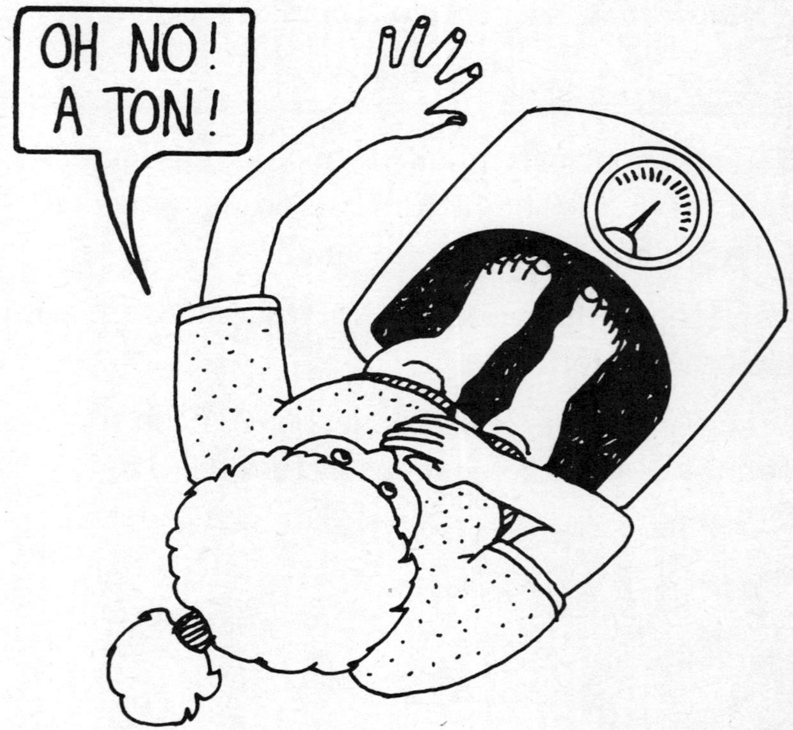

What are some things that are measured in tons? (Don't say "dads.") Ask your parent to think of two more things.

LENGTH X WIDTH = FUN

You need: a number cube, 2 different colored crayons

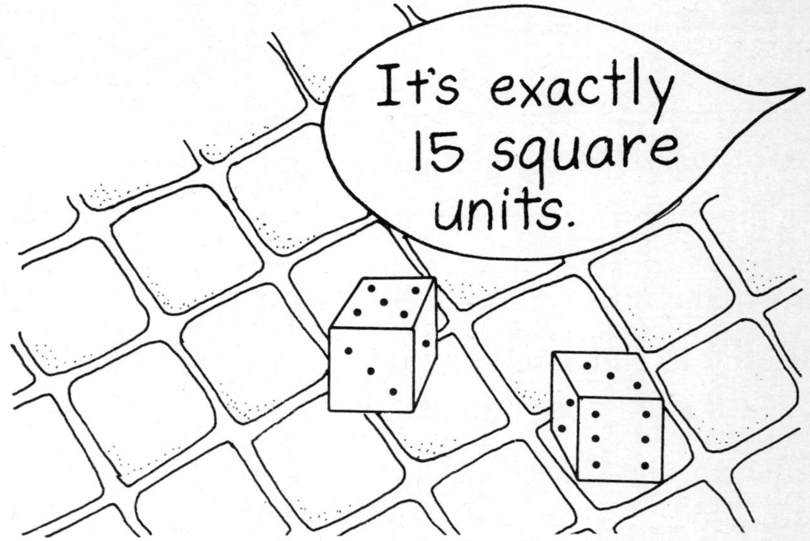

Take turns with your parent.

Roll the number cube twice and pretend the numbers rolled are the length of each side of a rectangle.

Find the area of that rectangle.

With your crayon, color the square on the gameboard that matches the area you found.

If you roll two numbers that give an area already colored, you're out of luck—lose a turn.

The first player to color three in a row in any direction is the winner.

$$A = l \times w$$

Area is equal to length times width.

1	2	3	4	5	6
7	8	9	10	11	12
13	14	15	16	17	18
19	20	21	22	23	24
25	26	27	28	29	30
31	32	33	34	35	36

BOX IT IN

You need: paper and pencil, tape, a cm ruler

Trace the illustration.

Then fold and assemble with tape to make a cube. (You'll need your parent's help here.)

Measure the sides and figure the volume.

Write it here: _____ cm^3

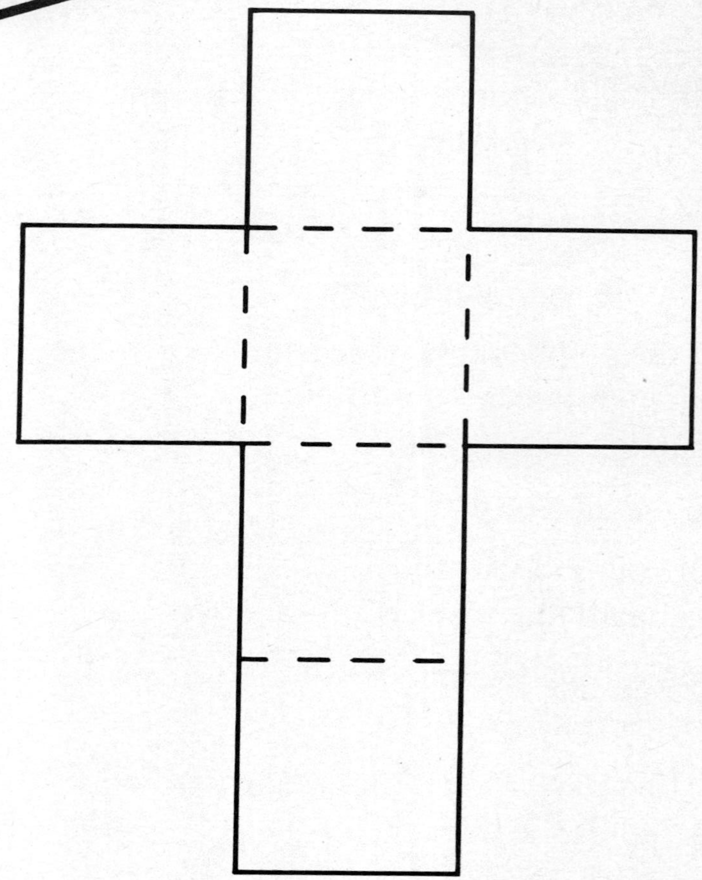

With your parent's help, try to make a paper box 3 cm × 4 cm × 3 cm.

ALL BOXED IN

You need: 5 boxes, a cm ruler, a calculator (if you have one)

With your parent, collect 5 boxes, empty or full (of anything), and put them on the table. Then, do this:

- Estimate which boxes you think have the greatest volume.
- Put them in order, according to your estimate, from smallest to largest.
- Now compute the volume of each box by multiplying the length by the width by the height. (Round off to even centimeters, if you have no calculator.)
- Reorder the boxes if your original estimates were wrong on any box.

A Postal Excursion

Go on a trip to a post office with your parent. Take this page with you and record the answers to the questions below. Look for answers or ask the postal worker to supply answers you cannot find by yourself.

Find:

The cost of a first class stamp	
The cost of a stamp for a post card	
The number of stamps in a book of stamps	
The cost of an air mail stamp to __________ (your favorite foreign country)	
The cost per ounce to mail a package, first class	
The cost per ounce to mail a package to __________ (your favorite foreign country)	

On the next two pages you'll chart and then summarize this information.

COMPARISONS, COMPARISONS (A Postal Excursion, page 2)

After you have gathered the information from your trip to the post office, complete these charts:

First Class Stamps	
Cost of 1 stamp	
Cost of 10 stamps	
Cost of 100 stamps	
Cost of 1000 stamps	

Post Cards	
Cost of 1 post card	
Cost of 12 post cards	
Cost of 35 post cards	
Cost of 82 post cards	

Packages	
Cost per ounce, 1st class	
Cost for 4 one-ounce packages, 1st class	
Cost for 40 one-ounce packages, 1st class	
Cost for 400 one-ounce packages, 1st class	

POSTAL PROBLEMS (A Postal Excursion, page 3)

Describe two things you learned while on your trip to the post office.

Write four story problems using the information you gathered on your trip.

BEAT THE CLOCK!

You need: a timer or clock

Ask your parent to set a timer or watch a clock. See if you can give the quotient for all the problems on one circle in 2 minutes. Write your answers inside the circle. Go! Stop! Did you beat the clock?

Then ask your parent to set the timer again while you try to give the quotient for all the problems on the other circle in 2 minutes. Did you beat the clock this time?

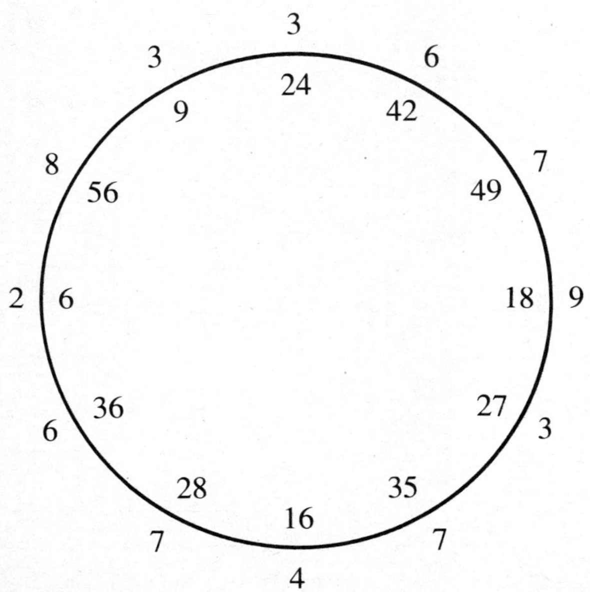

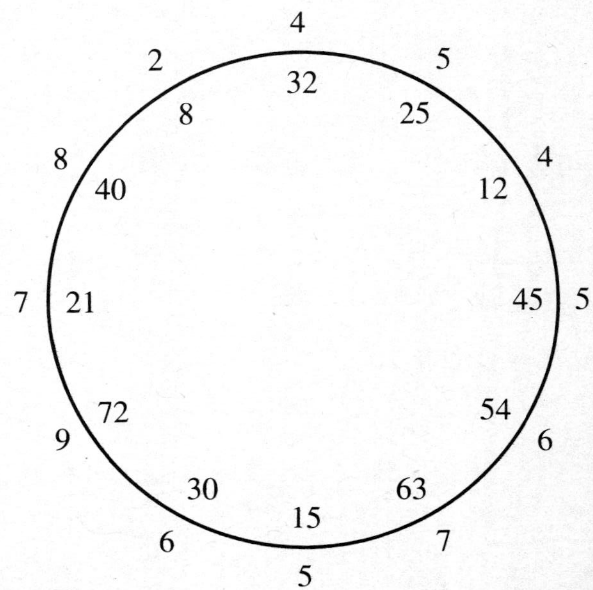

Division Rummy ÷

You need: 30 index cards or small pieces of paper

Shuffle the cards and deal 7 of them to each player. Stack the remaining cards face down. Check the cards you received for matching patterns: $1500 \div 3$, $150 \div 3$, $15 \div 3$. If you have 3 cards that match in this way, put them in a stack for 1 point. Each stack is called a book. The object is to try to make more books.

Draw a card.

- If the card matches one card you have, place the pair face up.
- If the card matches a pair you have, set the 3 cards that make a book aside and get a point.

Take turns drawing. Play until all the cards are turned up.

Whoever has the most "books" at the end is the winner.

Help your parent make these cards.

$35 \div 7$	$350 \div 7$	$3500 \div 7$
$32 \div 8$	$320 \div 8$	$3200 \div 8$
$16 \div 4$	$160 \div 4$	$1600 \div 4$
$42 \div 6$	$420 \div 6$	$4200 \div 6$
$18 \div 2$	$180 \div 2$	$1800 \div 2$
$2400 \div 8$	$240 \div 8$	$24 \div 8$
$3000 \div 5$	$300 \div 5$	$30 \div 5$
$1500 \div 3$	$150 \div 3$	$15 \div 3$
$4900 \div 7$	$490 \div 7$	$49 \div 7$
$5400 \div 9$	$540 \div 9$	$54 \div 9$

Quotient Toss

Write the numbers 2 – 9 on the beans.

Place the beans in a bowl. Lay your book on a table open to this page. Draw a bean from the bowl and toss it onto the game board in the picture.

Divide the number on the game board by the number on the bean. (Example: Toss a 5 bean. It lands on 32. The quotient is 6 with a remainder of 2). The *remainder* is your score for that round.

Take turns with your parent. First player to 35 wins.

You need: 8 beans, bowl

32	9	45	24
27	15	54	72
64	48	18	63
56	12	49	21
16	28	81	42

What number combinations would a good bean tosser try for? Why?

I Guess, About . . .

Take turns playing this division estimation game with your parent.

- Pick a problem from the box.
- Estimate the quotient.
- Check your estimate. (You may use paper and pencil or a calculator.)
- Use the key to find your score.

Cross out each problem after you've done it.

After 5 turns, total your scores to see who has the higher.

If you have a remainder, score the points PLUS the remainder.

PROBLEM BOX				
$6\overline{)96}$	$3\overline{)69}$	$4\overline{)96}$	$3\overline{)78}$	$9\overline{)99}$
$7\overline{)88}$	$5\overline{)85}$	$8\overline{)96}$	$6\overline{)72}$	$4\overline{)87}$

Score Key	
If your quotient is between:	
10 and 15 score	5 points
16 and 20	4 points
21 and 25	3 points
26 and 30	2 points

DIVIDE TO THE FINISH

You need: a paper clip, a pencil, different coins, 52 paper squares numbered 9 – 60.

Help your parent make the number cards. Hold the paper clip in place on the spinner with your pencil point.

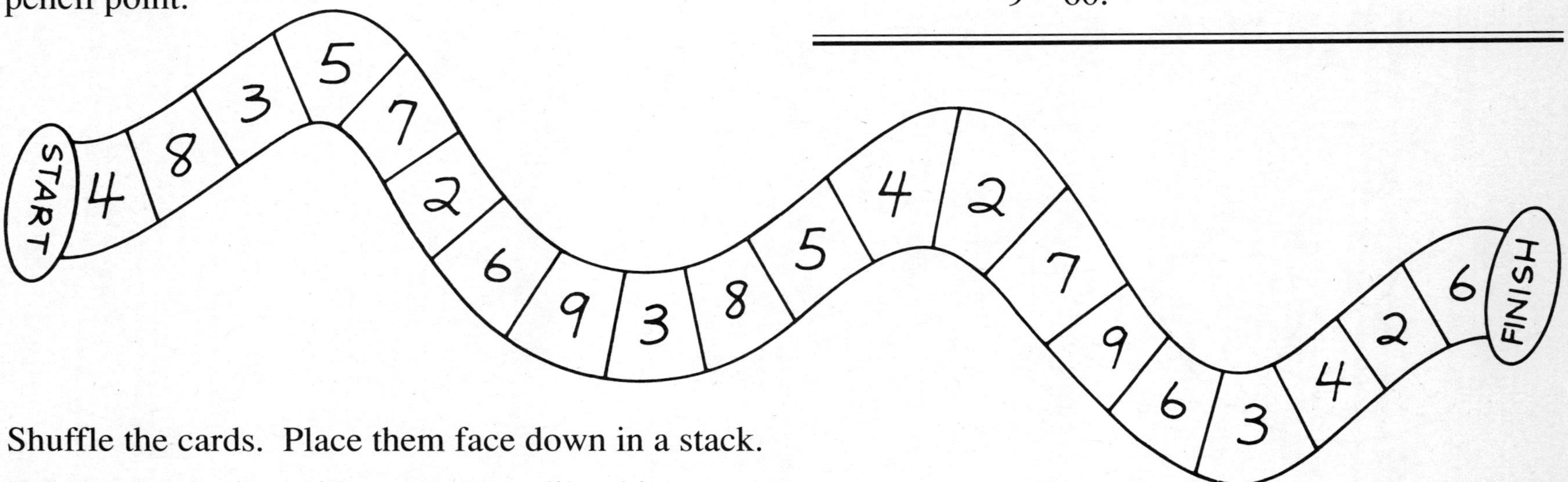

Shuffle the cards. Place them face down in a stack.

Take turns playing with your parent like this:

- Spin the spinner. Move your coin that many spaces.
- Draw a card.
- Divide the number on the card by the number on the space. Tell the quotient and the remainder.
- If correct, you may stay on the number. If incorrect, go back to your old space.

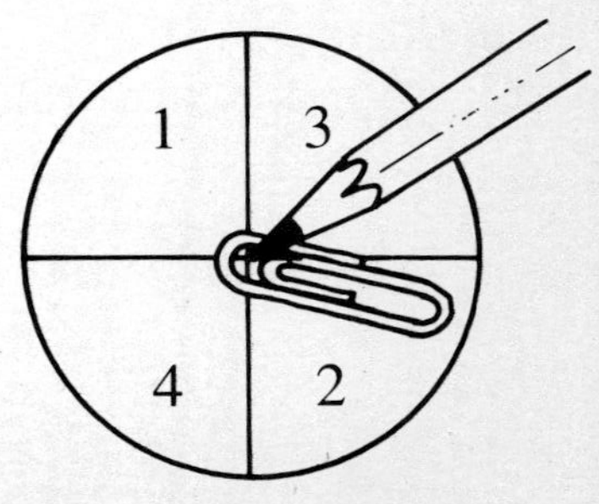

THE WHEEL OF MISSING FACTORS

Play with your parent.

Each of you choose a number between 2 and 9. Write the number in the center of your wheel.

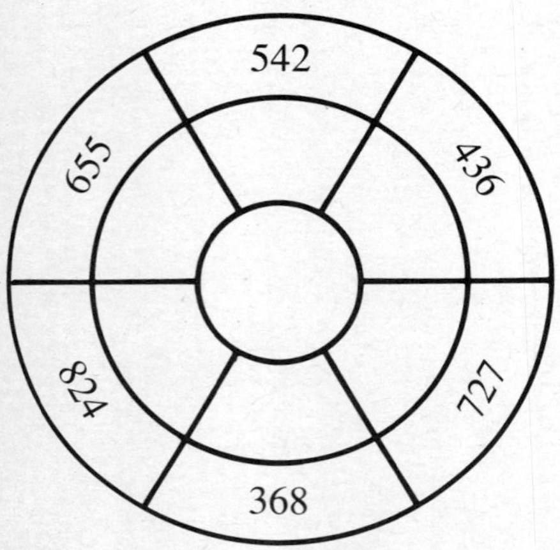

My Wheel

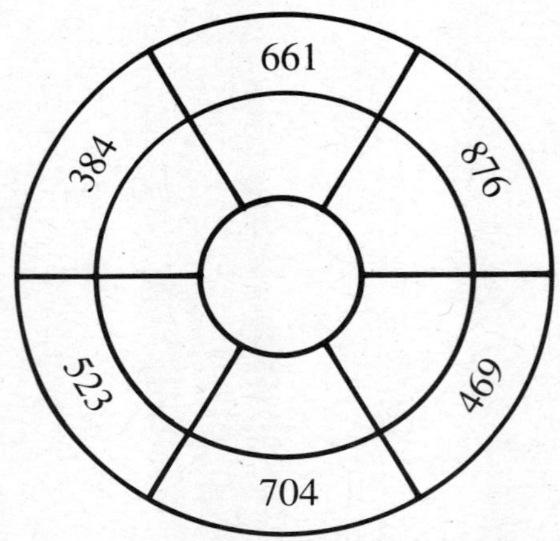

My Parent's Wheel

Take turns playing:

- Divide each number in the outer ring by the number in the center.
- Write the missing factor in the inner ring. Show remainders with *R*.
- See who can finish first.

It Averages Out

Find these numbers for yourself, your parent, and 2 or 3 other people.

- number of months spent in school since kindergarten (a full school year is 9 months)
- number of letters in full name
- number of members of extended family (including grandparents, aunts, uncles, and first cousins)
- age in months
- height in inches

Put the numbers on the chart.

Then find the averages in every column.

Name	*Months in School*	*Number of Letters in Name*	*Number of Family Members*	*Age*	*Height*
Total					
Average					

Zero Calculation

You need: 2 crayons

You and your parent each take a game board. Take turns playing.

- Select a division circle and solve the problem.
- Mark the circle off with an X.
- Find the quotient on your card and color in the box.
- The first player to color 4 consecutive blocks across, up and down, or diagonally, wins.

Game Board 1

30 R3	$2.08	104	104 R4
$1.02	106	203 R3	$2.04 R1
104 R3	30 R1	160 R1	105
104 R4	$1.20 R1	102 R6	30 R1

Game Board 2

102 R6	$2.30 R1	104 R4	$1.02
105	$2.04 R1	106	203 R3
104 R4	160 R1	30 R1	$2.08
104	$1.20 R1	30 R3	104 R3

3)91 5)523 6)624 4)815 3)$6.24 9)924 7)735 9)273 9)$9.18 4)641 8)848 5)524 8)836 7)$8.41 4)$8.17

Chinese Checkers Division

Make the cards shown with your parent.

Shuffle the cards. Turn them face down.
- Draw a card and find the quotient.
- Color in any block with your crayon.
- Then your parent has a turn and colors in a block with his or her crayon.

You need: division cards, 2 colored crayons or markers

Whoever gets a path of blocks connecting 2 sides of the game board (a minimum of 6 blocks) wins. You may color in a block to blockade your parent's path.

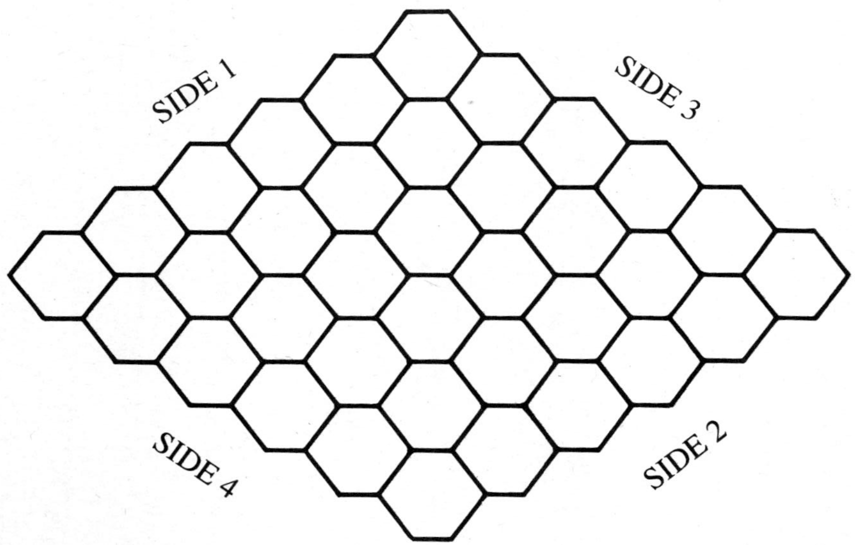

3)1946	6)3267	9)2841	7)5128
2)1878	5)3381	4)6025	8)2943
2)21,108	5)25,480	8)32,488	4)16,484
7)14,217	9)27,915	3)42,263	6)36,422
5)30,144	4)34,902	6)37,604	9)63,927

Handspans!

Ask the people in your house if you may measure their handspans. Measure in centimeters.

Record the measurements you make on the chart below.

Name	*Handspan in cm*

You need: a cm ruler, a pencil

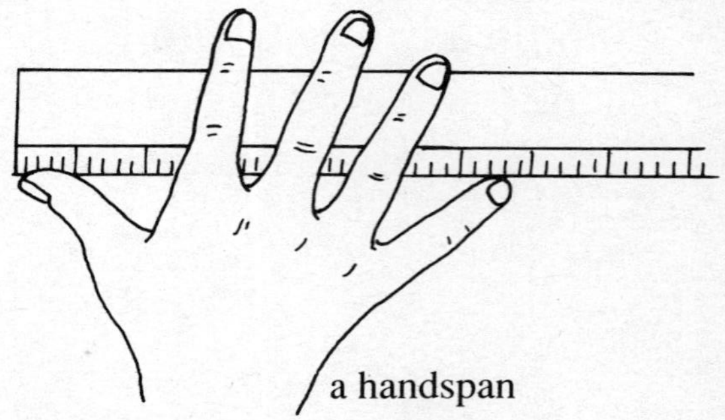

a handspan

To find the average handspan of the people in your house, fill in the boxes below.

The sum of all the people's handspans ↓		The number of hands measured ↓		Average handspan ↓
[ ]	÷	[ ]	=	[ ]

Discuss with your parent the range of your data (the difference between the longest and shortest measurement).

HIGH JUMPERS

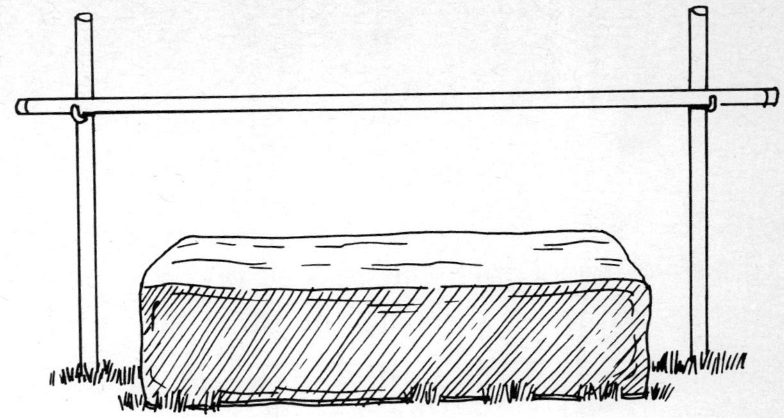

You need: a pencil, colored pencils

Rebecca's five highest jumps were:

1. 4′ 4″
2. 4′ 8″
3. 4′ 7″
4. 4′ 9″
5. 4′ 2″

Enter the height of Rebecca's jumps in the boxes below, beginning with the lowest jump.

lowest ☐, ☐, ☐, ☐, ☐ highest

Discuss the next two questions and their answers with your parent.

- What is the average height of Rebecca's jumps?
- What is the median (middle) height?

How high can you jump?

Draw yourself jumping over the bar.

Frisbee Fun

You need: a pencil, a Frisbee (if you have one)

These young people made some pretty decent Frisbee throws at the park the other day.

Names	*Frisbee Throws Measured in Strides*	*Mean*	*Median*	*Range*
Teresa	14.5, 15.0, 15.5 strides			
Sandy	12.8, 13.0, 14.5 strides			
Kenneth	16.0, 17.5, 18.5 strides			
Elmer	15.5, 17.5, 19.0 strides			

Ask your parent to help you calculate or choose, for each young person's Frisbee throws, the

- mean (the average measurement)
- median (the middle measurement)
- range (the difference between the longest and the shortest measurement)

Enter your own name and data for 3 of your Frisbee throws in the space provided on the chart. If you don't have a Frisbee, throw something else.

Extended Activity

Packages, Boxes, and Cartons

Take a little trip with your parent to a supermarket or a grocery store. Take this sheet of paper with you and on this form list at least 8 items that are sold in packages, boxes, or cartons containing more than one item. Write down the number of items in the package and its cost.

(Complete this column when you get home.)

Item	*Number of Items in Package, Box, or Carton*	*Cost of Package, Box, or Carton*	*Cost of Each Item in Package, Box, or Carton*

Ask your parent to keep the cash register receipt. Divide the cost of the package by the number of items in the package to find the cost of each item.

THE REAL COSTS (Packages, Boxes, and Cartons, page 2)

When you get home from the supermarket, do the following:

Find the cost per item of the items that came in packages, boxes, or cartons. Write the amount per item on the *first page* of this activity in the right column.

List some items that your parent bought, on this form.

Items Purchased	*Cost of Items*	*Number of Ounces*	*Cost per Ounce*

Examine the items your parent bought. Read the labels on the packages. Record the weight in ounces. Divide the cost of the items by the number of ounces to find the cost per ounce.

SMART SHOPPING — A STORY (Packages, Boxes, and Cartons, page 3)

Write a summary of the calculations you made. Explain how this kind of calculation can help you find better buys on items at the supermarket. Particularly, tell how knowing these things can tell you whether it is smarter to buy the smaller size of a product or the larger size.

Which Way Is Better?

Your parent reads aloud for you each of the story problems. You tell 2 different ways to solve each problem.

A. Mrs. Rivers had $20.00 when she went shopping. She spent $6.98 for a belt and $3.88 for a pair of socks. How much did she have left?

B. In a school election, 847 students voted. 284 voted for Tiffany, 269 voted for Morris, and the rest voted for Luke. Who won the election? How many votes did the winner receive?

C. Together Sylvia and Joan spent $3.00 for stickers and $5.00 for school supplies. If they split the cost equally, how much did each girl spend?

D. A sculptor bought 340 pounds of clay in a month. He bought 70 pounds per week for 3 weeks. How many pounds did he buy the fourth week?

With your parent, solve 2 of the problems both ways. Are the answers the same?

Who Estimates Better?

You need: 2 paper clips, a pencil, a calculator if you have one

Play the game with your parent. Hold the paper clips down with your pencil point.

- Spin each spinner.
- Try to guess the quotient of the numbers you spun.

Here's how:

Without pencil and paper, round the 2-digit divisor to the nearest ten and the 3-digit dividend to the nearest hundred.

- Divide in your mind.
- Write your estimate on a sheet of paper.
- Use a calculator or pencil and paper, then, to find the actual quotient.

Each player's score is the difference between his or her guess and the correct quotient.

The *lower* score wins!

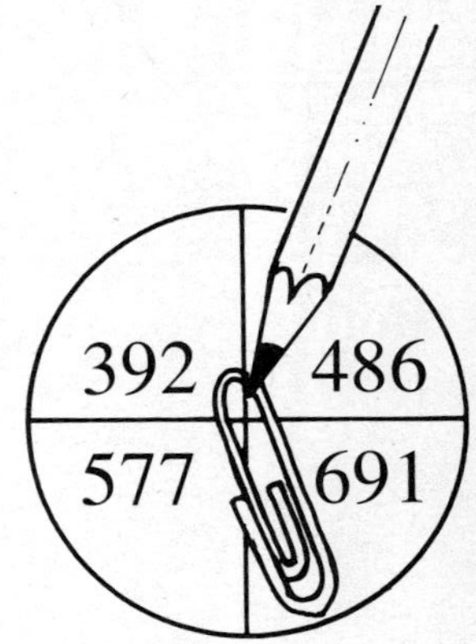

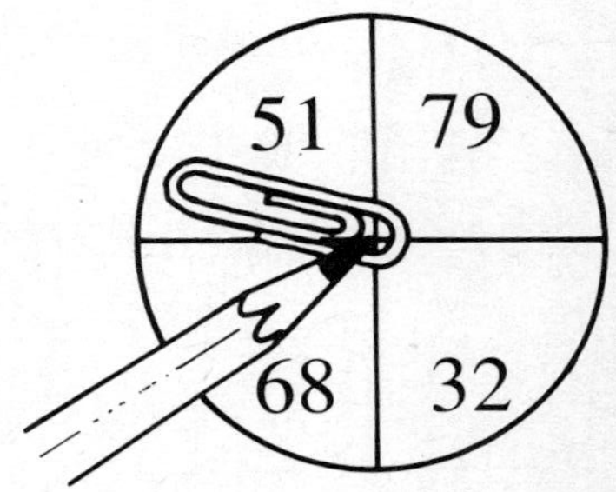

Why does the lower score win?

A FUN PLACE!

You need: 8 cards or paper squares

Write the division problems on the cards neatly. Shuffle the cards. Give half of them to your parent. Both of you find the quotients of 4 problems.

Arrange the quotients in a row from smallest to largest.

Fill in the blanks below with the letters on the cards as they now stand, to find a fun place for a vacation.

____ ____ ____ ____ ____ ____ ____ ____

O $87\overline{)522}$	D $58\overline{)522}$
L $52\overline{)416}$	W $76\overline{)380}$
A $96\overline{)384}$	E $65\overline{)195}$
R $89\overline{)623}$	S $59\overline{)118}$

Am I Correct?

You need: a bean, a red and a blue crayon

Drop a bean on the gameboard.

See what problem it falls on.

Estimate for your parent what you think the quotient will be.

Your parent divides either with pencil and paper or with a calculator to find the actual quotient.

If your parent says you were very close, color the block red.

If your parent says you were far off, color that block blue (for tears).

Did your red blocks far outnumber your blue ones?

$48\overline{)149}$	$23\overline{)189}$	$71\overline{)488}$
$32\overline{)317}$	$81\overline{)646}$	$62\overline{)423}$
$19\overline{)188}$	$52\overline{)561}$	$55\overline{)309}$

TIME - TO - DIVIDE

You need: a watch, pencil and paper

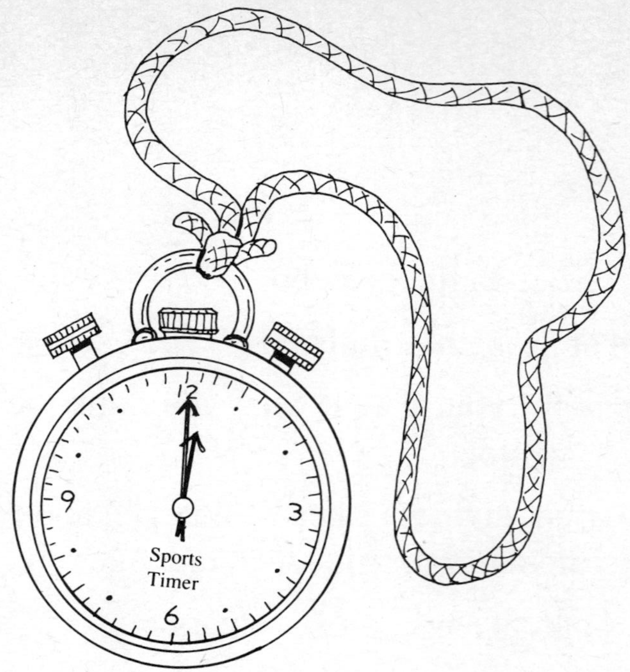

Make up 4-digit numbers for each dividend in Row 1. ($\overline{)1234}$)

Ask you parent to insert 2-digit numbers for divisors. (56$\overline{)1234}$)

Then your parent uses a watch to see how long it takes you to find the quotients and remainders for the problems in Row 1.

Work on scrap paper.

Row 1	$\overline{)}$	$\overline{)}$	$\overline{)}$	$\overline{)}$
Row 2	$\overline{)}$	$\overline{)}$	$\overline{)}$	$\overline{)}$

Play again, using Row 2. This time your parent makes up the 4-digit numbers for the dividends.

You make up the 2-digit divisors.

Ask your parent to time you again. Did your time improve?

Which Football Team?

Discuss with your parent the method you have learned in school for doing hard long division problems.

An example (calculating $16\overline{)3,915}$) is shown at right.

Calculate the answers to the following problems:

$29\overline{)5,510}$ $33\overline{)6,435}$

$42\overline{)6,090}$ $34\overline{)4,080}$

Now match the answers to the letter codes below. The name of a famous football team can be formed from the letters, if you unscramble them.

A	B	C	D	E	F	G	H	I	J	K	L	M
100	105	110	115	120	125	130	135	140	145	150	155	160
N	O	P	Q	R	S	T	U	V	W	X	Y	Z
165	170	175	180	185	190	195	200	205	210	215	220	225

The name of the football team is the: ____________

A method for calculating $16\overline{)3,915}$		
Stage 1	Estimate	200 $20\overline{)4,000}$
Stage 2	Divide the hundreds	2 $16\overline{)3,915}$ 3 2 7
Stage 3	Divide the tens	24 $16\overline{)3,915}$ 3 2 71 64 7
Stage 4	Divide the ones	244R11 $16\overline{)3,915}$ 3 2 71 64 75 64 11
Stage 5	Check with estimated quotient to see if it is reasonable.	

What's the Equation?

Ask your parent to help you make up 4 story problems, one each in addition, subtraction, multiplication, and division.

Addition: Use any number of pantry, refrigerator, and freezer items to compile a shopping list and total.

Subtraction: Find two similar items in the house (deodorant, hairspray, toothpaste, whatever). Find the difference in their weight or cost.

Multiplication: Use a simple recipe and increase it by four. (You're having the whole team over for dinner.)

Division: Find the cost per ounce in a package of food like beans or rice.

You need: 2 sheets of paper cut in half

Discuss each problem with your parent. If necessary, write the equation needed to solve the problem on a paper.

Solve the problem.

Extended Activity

A Mall Ball

Go on a trip to a mall with your parent. Take the charts on this page with you and complete them. Write the names of the stores you visited, what items you would like to buy if you were fabulously rich, and the items your parent would like to buy. Use another sheet of paper if you need to.

Stores we visited

Things I'd like to buy	*Cost*

Total cost: ________

Things my parent would like to buy	*Cost*

Total cost: ________

(Do these totals after you get home.)

The next two pages are about how to pay for all this stuff.

PAYING THE PIPER (A Mall Ball, page 2)

Do the tasks on this page after you get home from the mall.

Suppose you could buy all the items and pay for the total amount in 12 equal payments. How much would each payment be?

Do the same calculations for your parent's list.

Use this page to show your work.

ALL ABOUT THE MALL (A Mall Ball, page 3)

Talk to your parent about making installment payments to purchase expensive items like cars.

Ask about the advantages and disadvantages of buying this way.

Write a brief summary of your trip to the mall. Tell, also, what you learned from your parent about installment buying.

Part 3

Introducing Geometry: angles and spaces that shape our lives

A note to parents

Have you looked closely at the shirts children are wearing—or would like to wear—today? Along with the bold colors you've seen wild designs using geometric shapes such as angles, triangles, circles, and squares. That means geometry is a natural item in your child's life, even before you begin the activities in Part 3. Along with geometry, this part contains practice with decimals.

Try some of these decimals activities together

- Work together with real-life cases of multiplying or dividing decimals. For example, ask, "If I used 15.4 gallons of gas in the car this week, how much will I use in a month?"
- Talk about money amounts in terms of tenths and then in terms of hundredths. For instance, discuss $6.30 worth of dimes to help your child see tenths clearly, and then $6.30 worth of pennies.

Then help your child become familiar with geometric shapes.

- With your child, find everyday objects that have geometric shapes. Hunt for squares one day, circles the next, cones the following day. You both will be surprised by how many objects you find!
- Help your child find examples of right angles (90 degrees), acute angles (smaller than right angles), and obtuse angles (larger than right angles) at home. For example, the corner of a book is a right angle, and two panes of glass in a bay window form an obtuse angle.
- Experiment with mirror reflections of various shapes. Have your child identify which shapes are mirror images of each other, and introduce the term *symmetry*.

Then try the activities in Part 3 together.

PAY YOUR WAY

You need: a pencil, colored pencils

High School	College	Professional
________________	________________	________________
GAME TODAY!	GAME TODAY!	GAME TODAY!
TICKETS $3.50	TICKETS $15.50	TICKETS $25.50
Cost for 5 people ☐	Cost for 5 people ☐	Cost for 5 people ☐

Enter the name of your favorite team on the dotted lines. Decorate the posters.
Calculate the cost for each game if you were paying for yourself and 4 friends.
Enter the amount in the 3 boxes and discuss your calculations with your parent.

Why are the costs for each game so different?

A Neat Cereal Box

Imagine you have been asked to design the front of a cereal box.

Talk with your parent about the following instructions:

- Invent a name for the new cereal.
- Think of a picture that goes well with the name. Draw it.
- Convert the ounces into grams (because the new cereal is also intended for the overseas market). Remember, 1 oz. = 28.35g.
- Check the numbers for ounces and grams on one other cereal box in your cupboard.

You need: a pencil, colored pencils

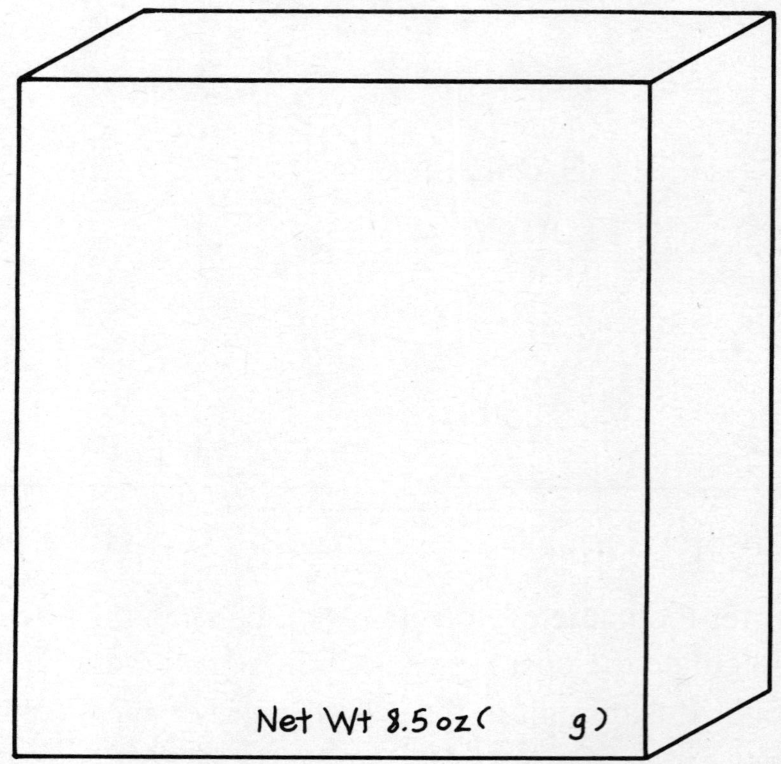

How FAR Will It Go?

Finish this drawing of a bicycle.

Then ask your parent to help you accurately figure out the circumference of one of these bike tires (that is, the distance *all* the way around one). (In case you forgot, the formula is $C = \pi d$.)

If the bicycle wheel turns 20.5 times, what distance would this tiny bike travel? Ask your parent to help you with the calculation.

You need: a tape measure, a pencil

Use 3.14 for π.

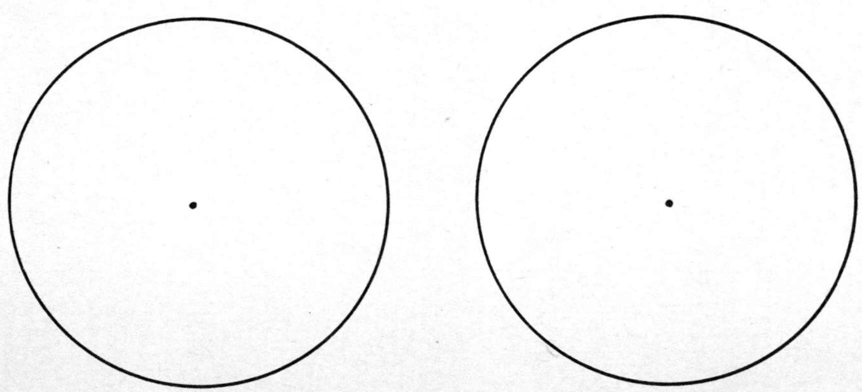

Why does the speed you travel make no difference to the distance covered?

Pounds to Dollars

The Washington family are in England for a vacation. The children, Mary and Greg, enjoy riding the London buses and taxis. They have found cute models of the bus and the taxi in a toy store. The children have decided to buy one each as a souvenir. Naturally, the price of the toys is written in pounds. Mary and Greg need to know how much the toys would cost in dollars.

Help these two youngsters change pounds to dollars! Round each to the nearest hundredth.

Ask your parent to help you calculate the cost in dollars for each toy. On that day, the exchange rate was: £1 = $1.58.

You need: a pencil, colored pencils

Mary chose a London taxi

Cost in pounds £3.49

Cost in dollars $

Greg chose a London bus

Cost in pounds £2.99

Cost in dollars $

Find a picture of a double-decker London bus and draw it in the box.

HOW FAR?

You need: a cm ruler, a pencil

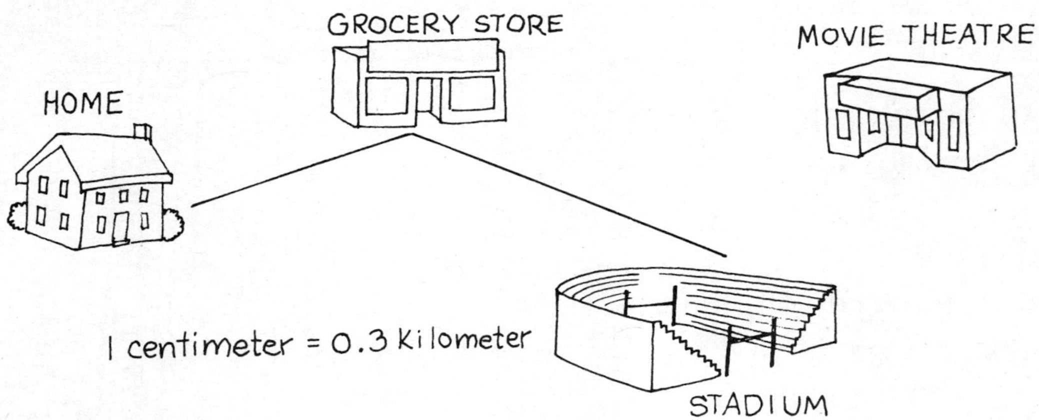

Measure the distances in the tiny map in centimeters and enter your data in the boxes.

Ask your parent to help you calculate the actual distances, using the scale on the map (1 cm = 0.3 km).

			Actual distance	
• Home to the Grocery Store =	☐	centimeters	☐	kilometers
• Home to the Stadium =	☐	centimeters	☐	kilometers
• Home to the Movie Theatre =	☐	centimeters	☐	kilometers

Enter the data in the appropriate boxes and discuss the results with your parent.

Quick Reactions

You need: a cm ruler, a pencil

Ask your parent to hold a ruler vertically in the air with the zero mark downwards.

Hold your thumb and finger on either side of the ruler at the zero mark, without touching the ruler. Then you are ready to do the experiment:

- Your parent drops the ruler without warning you.
- As soon as your parent lets go, trap the ruler between your thumb and finger, and note the measurement. This measurement shows the distance travelled by the ruler during your reaction time.
- After one practice attempt, try 3 more catches, and enter your results in the boxes. Measure in centimeters.

☐	+	☐	+	☐	=	☐	÷ 3 =	☐
Catch 1		Catch 2		Catch 3		Total Distance		Average Distance

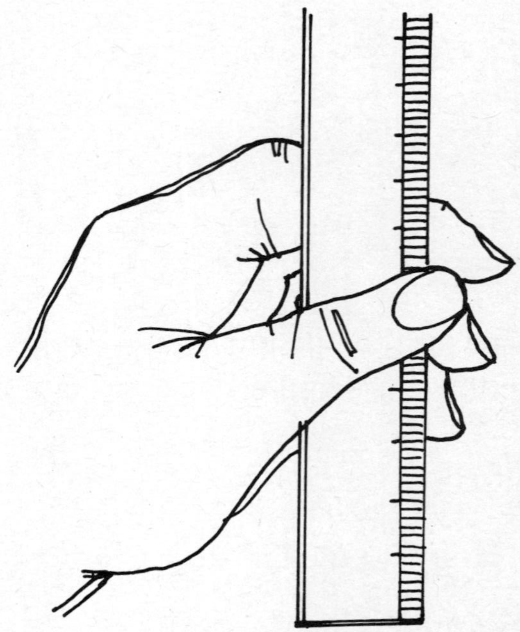

Switch places with your parent now, and compare reactions.

The Mystery Numbers

Ask your parent to watch you figure this one out.

You need: three 1" squares cut from paper or an index card, a pencil

$$4\overline{)3.1**} = 0.*88$$

Game rules:

- On your three 1" square cards, write a 2, a 5, and a 7.
- Begin to solve the problem of where to place the 2, 5, and 7 in the problem above.
- When you think you have completed the problem, ask your parent to test the accuracy of your solution. If it is correct, you're a winner. If not, the problem solving continues until a winning solution is found.

Ask your parent to make you another puzzle like this, or two.

Look for the Pattern! Pattern! Pattern! Pattern!

Look at this calculation with your parent, and discuss what might happen if you continued the calculation using the next six zeros.

```
    .1 4 2 8 5 7 1
7 ) 1.0 0 0 0 0 0 0 0 0 0 0 0 0 0
   -7
    3 0
   -2 8
      2 0
     -1 4
        6 0
       -5 6
          4 0
         -3 5
            5 0
           -4 9
              1 0
              -7
               3
```

Now continue the calculation and see if you were right. What pattern do you see?

Do patterns like this occur with other numbers?

Ten or a Hundred Times Bigger?

You need: 3 index cards, each one cut into quarters, and a pencil

On each card, write one of the following numbers:
0, 0, 0.2, 1, 2, 6, 9, 10, 15, 60, 90, 150
Shuffle the cards thoroughly and ask your parent to play the game with you.
Share the cards equally between you and your parent.

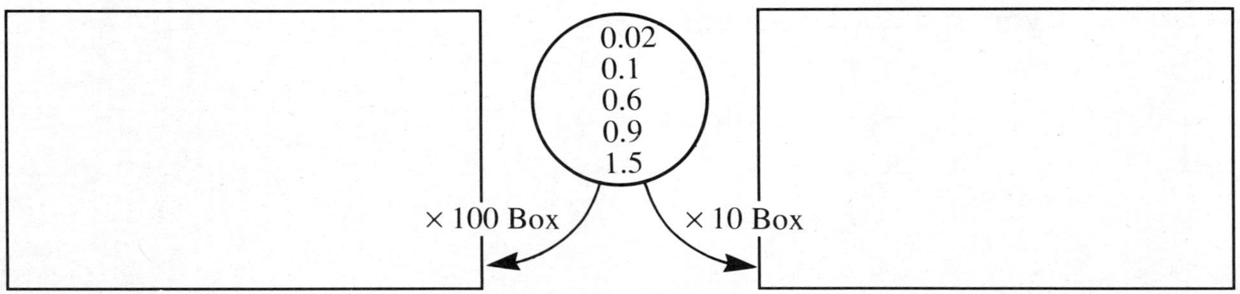

Scorecard		
Player 1	Player 2	
		Total

- Both of you write your initials on every one of your cards.
- Take turns placing a card in the correct area on the game board. (For example, to play the 150 card, it must be placed in the × 100 box)
- Drawing a 0 card misses a turn.
- When one player has placed all his or her cards, the game is over.
- Stop and enter all the numbers from the cards into the appropriate column in the scorecard.
- Then total the columns.

The winner is the player with the higher score.

How do you quickly multiply by 10 or 100?

One Thousand Times Bigger

You need: 2 number cubes

Game rules

- Choose a card; the other belongs to your parent.
- Each in turn shakes 2 number cubes, and rolls them.
- The number on the left cube will be used as tenths; the number on the right will be used as hundredths (see the picture).
- Multiply the number rolled by 1000. If your answer is correct, you may shade the appropriate number on your card.
- The winner is the player who first completes a vertical, horizontal, or diagonal line of shaded numbers.

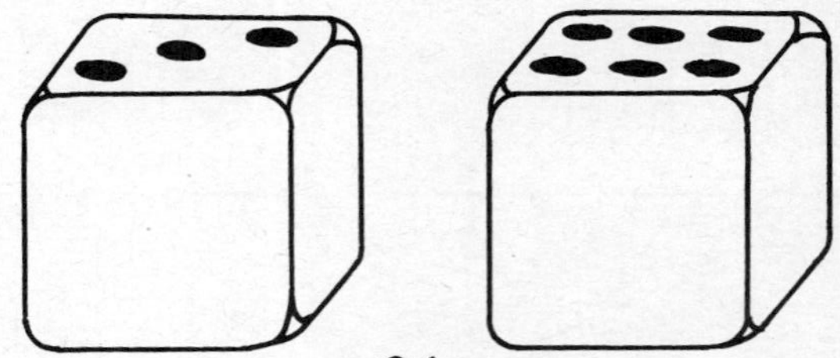

.36

Your Card

110	210	310	410	510	610
120	220	320	420	520	620
130	230	330	430	530	630
140	240	340	440	540	640
150	250	350	450	550	650
160	260	360	460	560	660

Parent's Card

110	210	310	410	510	610
120	220	320	420	520	620
130	230	330	430	530	630
140	240	340	440	540	640
150	250	350	450	550	650
160	260	360	460	560	660

Does this game depend on luck or skill?

Scavenger Hunt

You need: string, a centimeter ruler, a pencil

What to do:

1. See if you and your parent can locate 5 items in your house that are in the shape of a circle.
2. Measure the diameter of each item to the nearest cm and record it in the table.
3. Now, for the hard part—calculate the circumference of each item and record it.

Item	*Diameter*	*Circumference*

Remember: $C = \pi d$

C = circumference

$\pi = 3.14$

d = diameter

Now check your calculations by *measuring* all the circumferences using string.

Food Search

- Ask your parent to help you choose different-sized cans from your kitchen.
- Measure the diameter of each and record it in the box to the left. Find the radius (r).
- Multiply $3.14 \times r^2$ to find the area of each can top.
- Use a calculator if you have one or ask your parent to check your multiplication.

Item	*Diameter*	*Radius* ($\frac{1}{2}$ of the diameter)	*Area*
1.			
2.			
3.			
4.			
5.			

You need: 5 cans of any kind, a cm ruler, a calculator (if you have one)

Example:

The diameter is 3.4 cm so, radius is 1.7cm.

$$1.7^2 = 1.7 \times 1.7: \quad 119 + 170 = 2.89\ cm^2$$

$$2.89\ cm^2 \times 3.14: \quad 1156 + 2890 + 86700 = 9.0746\ cm^2$$

area = $9.0746\ cm^2$

Match a Measure

You need: a number cube, 2 different-colored crayons, your math book

- Take turns throwing the cube.
- Pick a box with the number thrown in the right-hand corner.
- Match one of those numbered boxes with *an equivalent* numbered box. If you found the true equivalent, color in both boxes.
- If you roll a number with no boxes left to color, your turn is skipped.
- The last one to color in a match WINS!

Note: You will need your math book to help you with the formulas.

1  5 cm, 3 cm	2  5 cm, 3 cm	3 volume = $18\ cm^3$	4 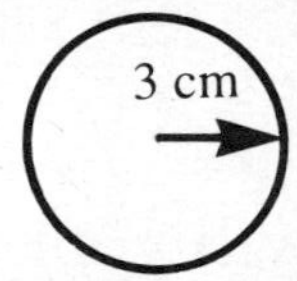 3 cm	5 circumference = 31.4 cm
6 area = $78.5\ cm^2$	1  7 cm	2 area = $15\ cm^2$	3 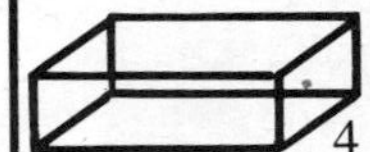 3 cm, 4 cm, 6 cm	4 diameter = 8 cm
5 circumference = 50.24 cm	6 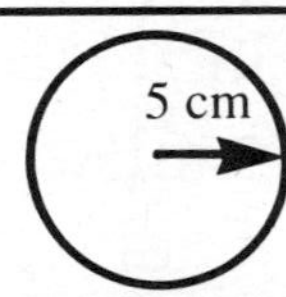 5 cm	1  2 cm	2 area = $9\ cm^2$	3  5 cm, 3 cm, 3 cm
4 diameter = 6 cm	5 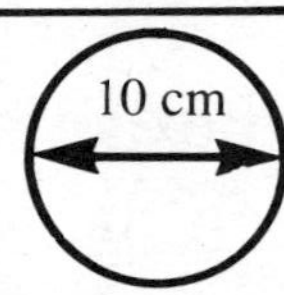 10 cm	FREE	1 perimeter = 16 cm	2 area = $4\ cm^2$
3 $45\ cm^3$	4  4 cm	5 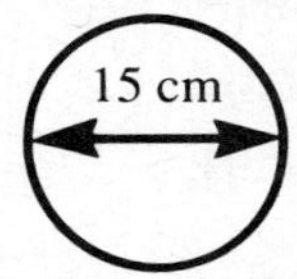 15 cm	6 area = $3.14\ cm^2$	1 perimeter = 8 cm
2  2 cm	3 volume = $72\ cm^3$	4 diameter = 100 cm	5 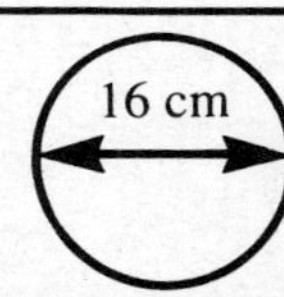 16 cm	6 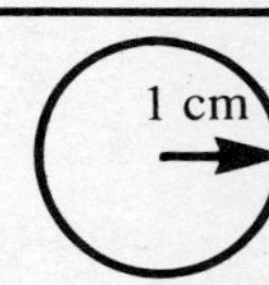 1 cm
1 perimeter = 21 cm	2 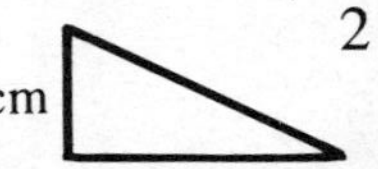 3 cm, 6 cm	3  3 cm, 2 cm, 3 cm	4 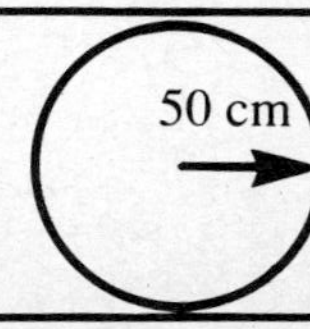 50 cm	5 circumference = 47.1 cm

GOOD MILEAGE?

How many miles will your car travel for each gallon of gasoline pumped into its fuel tank? Here is how to find out.

- With your parent's help, decide when the fuel gauge pointer has reached the $\frac{1}{4}$ mark. Read the odometer, then, including the tenths. The odometer will show you how many miles your car has traveled up to this point in time. Record your reading. (Keep a notepad and pencil in the car.)
- Each time gas is pumped into the car, find the exact amount of fuel which has been pumped, from the meter on the gas pump. Record the amounts on your notepad.
- To finish the test, record the mileage from the odometer as soon as the fuel tank pointer reaches the $\frac{1}{4}$ mark again.

The next page will show you what to do with the data you have collected.

GAS + MILEAGE FLOW CHART (Good Mileage? page 2)

Enter the data you have collected in the appropriate boxes.

Then sum up on the next page of this activity.

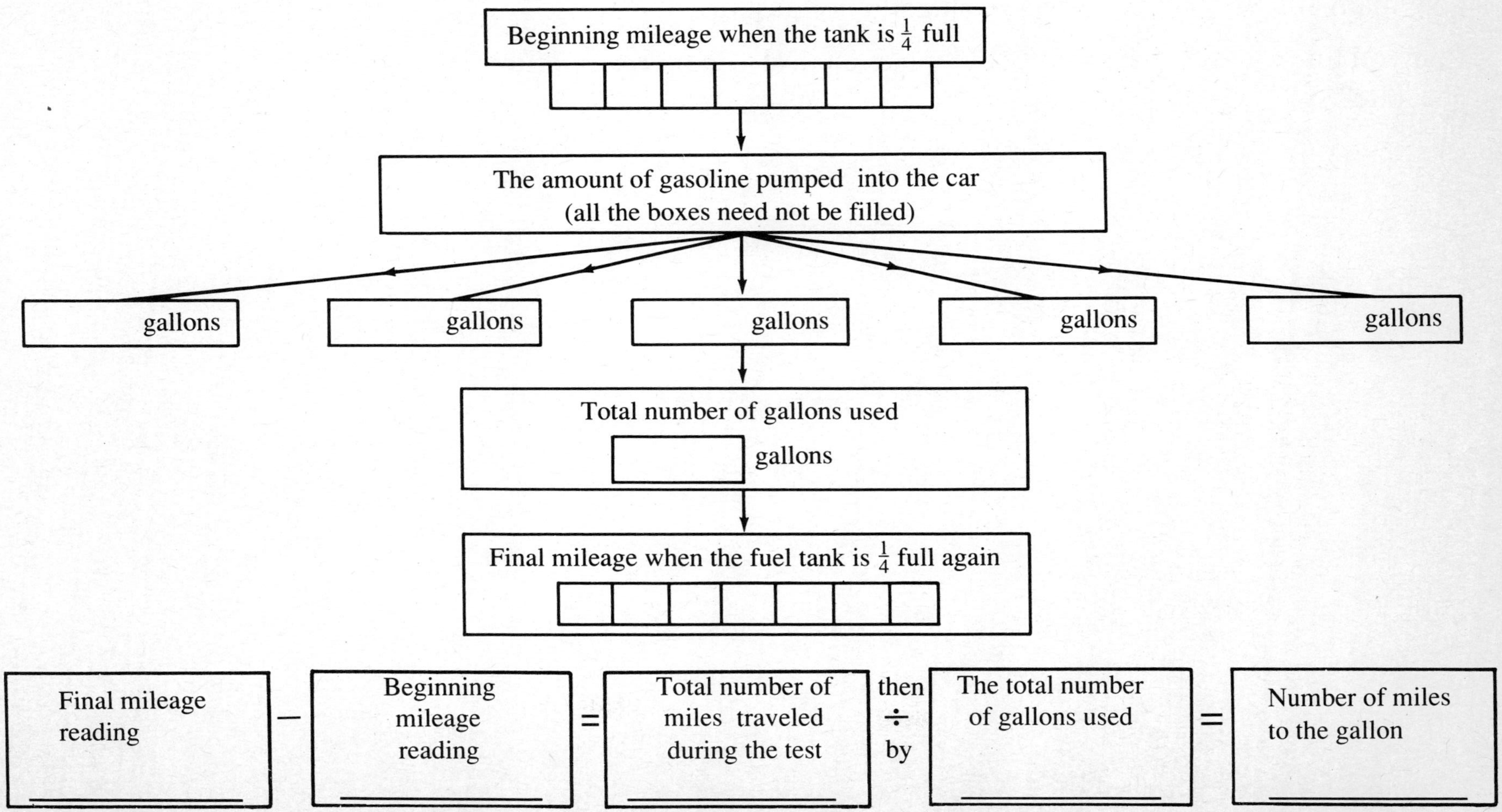

MILEAGE JUST THIS GOOD (Good Mileage?, page 3)

Talk to your parent about the result. Ask him or her to check your calculations. Were either of you surprised?

Draw a picture of your car, and underneath write a description of it, including how many miles it will travel for each gallon of gasoline.

Can you think of any ways you could make the test you carried out more accurate?

Match those parts!

- right angle
- acute angle
- obtuse angle
- ray
- center
- diameter
- chord
- CE
- C
- HG
- BF
- ACD
- FCE
- ACE

Label your 14 mini-cards as shown above. Then shuffle and place them face down in the shape of a circle.

Play this matching game with your parent, then, taking turns. To play:

- Turn over 2 cards and study the circle in the illustration to see if they match. If they do match, keep the cards. If not, put them back in the circle, face down.
- Continue the game until all the cards have been taken. The winner is the player with the most cards!

You need: 7 index cards cut in half , a pencil

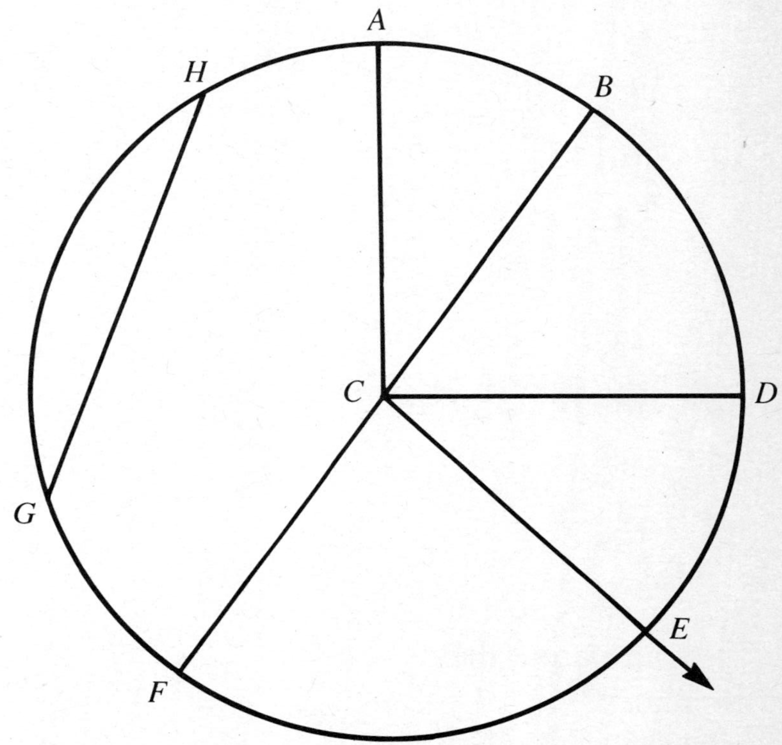

Why does the line continue after CE?

Chord 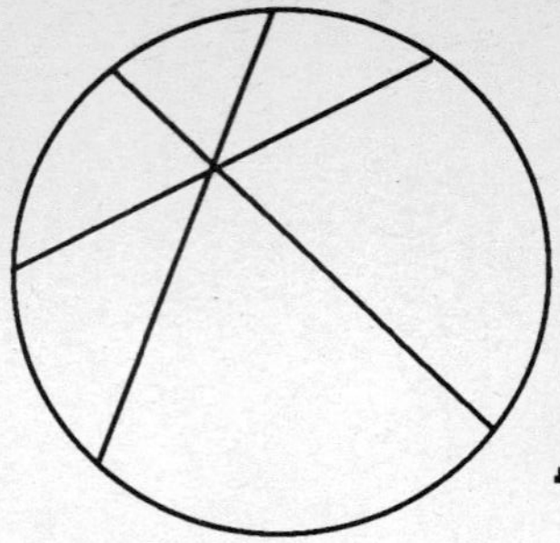 Art

You need: a ruler, a pencil

Use a ruler to draw the following chords:

1, 9	9, 17	17, 1
2, 10	10, 18	18, 2
3, 11	11, 19	19, 3
4, 12	12, 20	20, 4
5, 13	13, 21	21, 5
6, 14	14, 22	22, 6
7, 15	15, 23	23, 7
8, 16	16, 24	24, 8

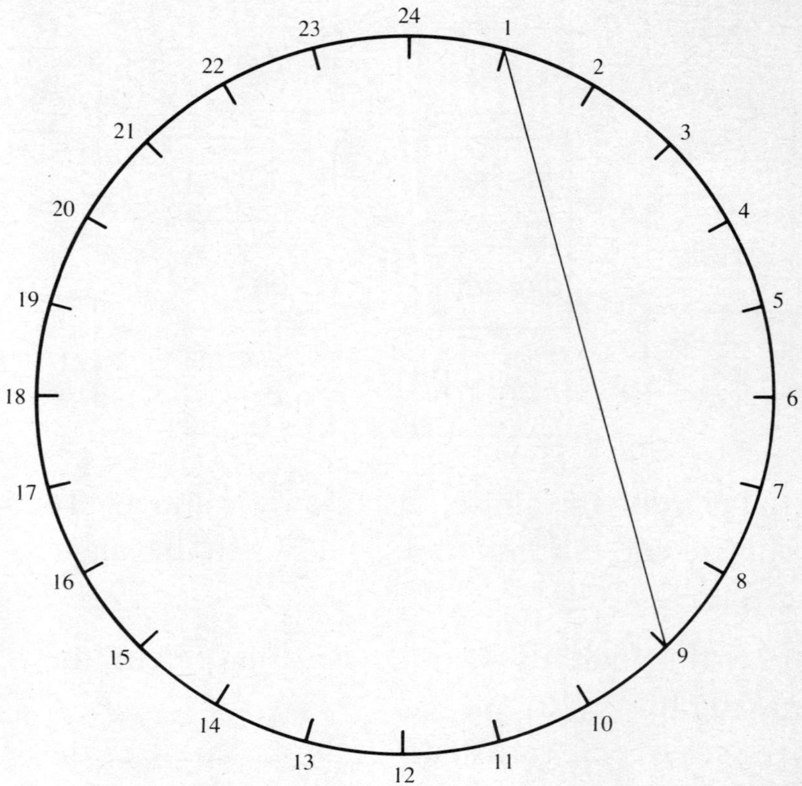

Show your parent the effect you have created.

This is a slightly different matter, but can you explain how a curved building can be built from flat pieces of glass?

How Many Degrees?

You need: a pencil, a protractor

1.
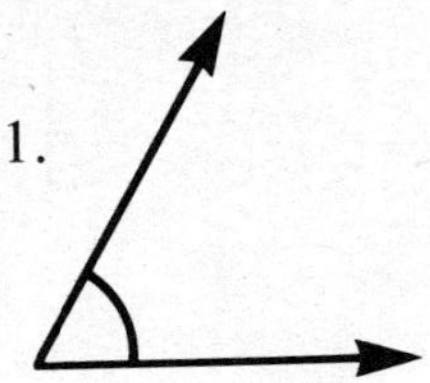

2.
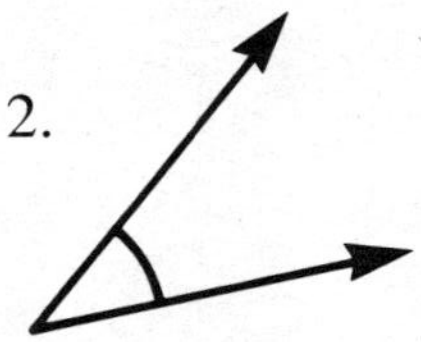

3.
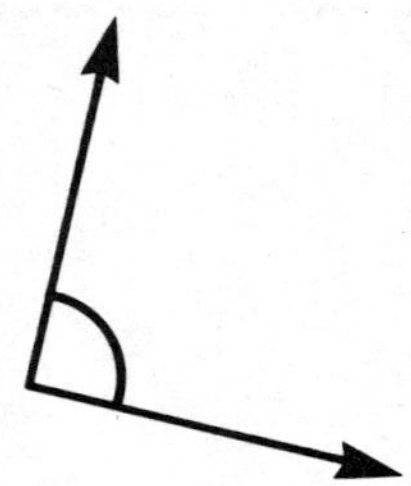

4.
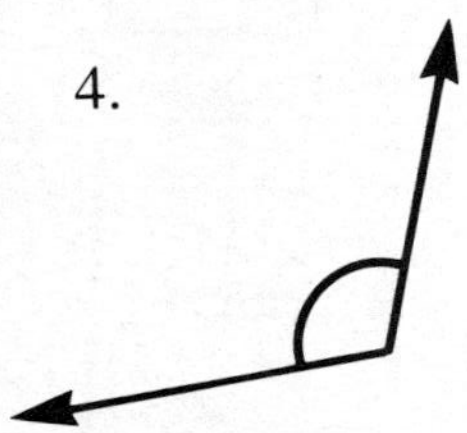

5.

6.

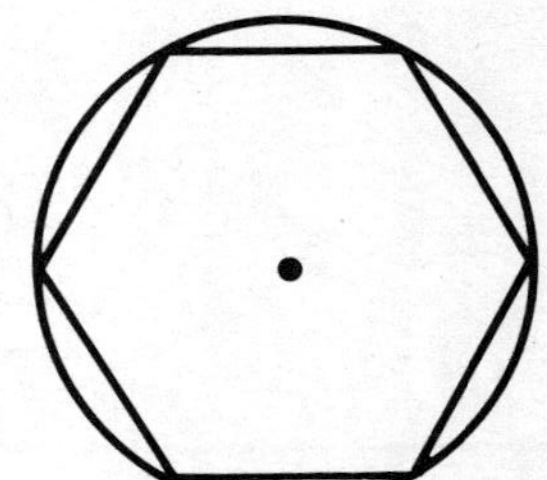
Me

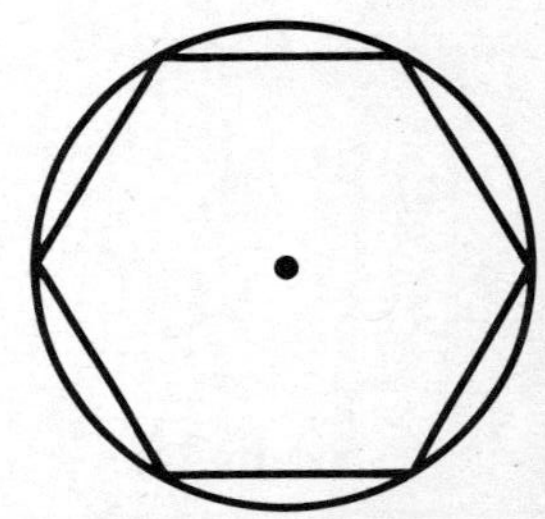
My Parent

- Ask your parent to join you in estimating the angles above.
- Measure the angles with a protractor, and decide whose estimate is nearer. The player with the nearer estimate draws a radius in one of the circles.
- The winner is the player who has drawn more radii.

Which was the most difficult angle to estimate? Why do you think this is?

Triangles in Construction

Talk to your parent about triangles you have both seen being used in various places.

bridge

roof

crane

bicycle

Draw pictures of a bridge, a roof, and a crane showing the triangles that have been used. Then think of a different example that you can draw in the last space.

Can you imagine a world without triangles? Talk with your parent about what it would be like.

How Many Triangles?

You and your parent will race to find all the triangles in this diagram! (Cover it up until your parent says "Start!")

You need: a pencil, a timer or watch, a protractor

1. For one minute, you and your parent each write down as many different triangles as you can. Especially look for congruent triangles! (Remember, congruent triangles fit over each other exactly.)
2. At the end of the minute, count up your score:
   - You earn *1 point* for each triangle you find.
   - You earn a bonus of *2 points* for each pair of congruent triangles you find. (That will be *4 points* in all—2 for two triangles, 2 more for their being congruent.)
3. Afterwards, circle all the pairs of congruent triangles.

Example: Triangles STU and WUV are congruent.

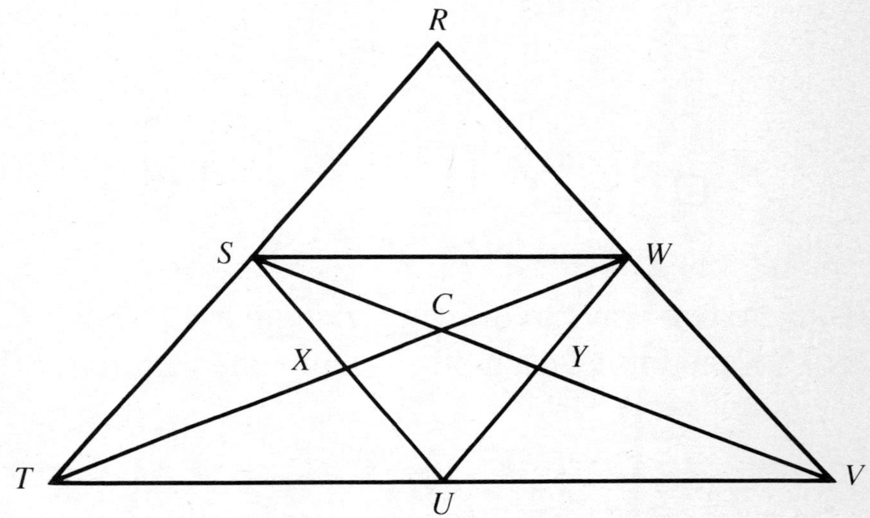

In the diagram, how many triangles are there altogether?

Squaring Triangles

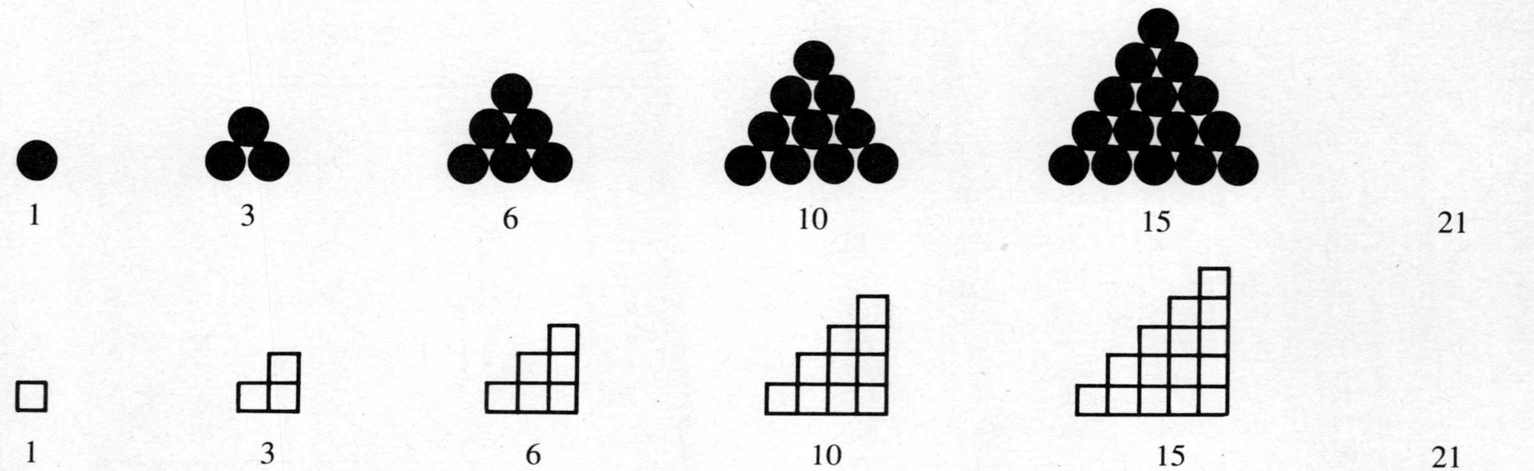

Here are two ways of drawing triangular numbers. Describe to your parent how the two sets of numbers are like and unlike each other. Draw the pattern for 21 in both sets.

See what happens when you turn the previous number in the series upside down and fit it into the shape of the number on its right?

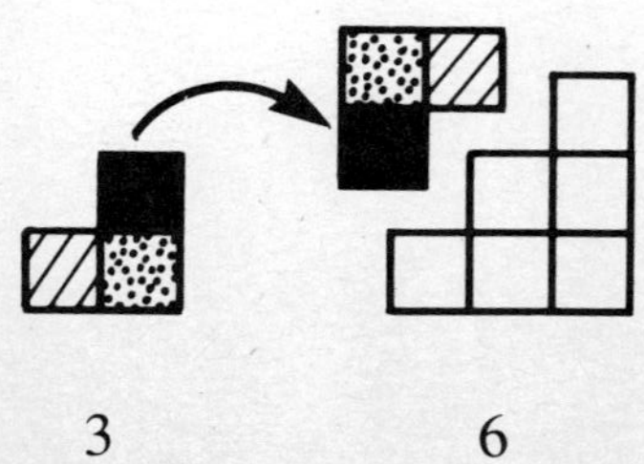

What kind of a number do you have now? Try it with other triangular numbers.

Give Me 5!

You need: scissors; tape or glue; paper

Sample Pentominoes

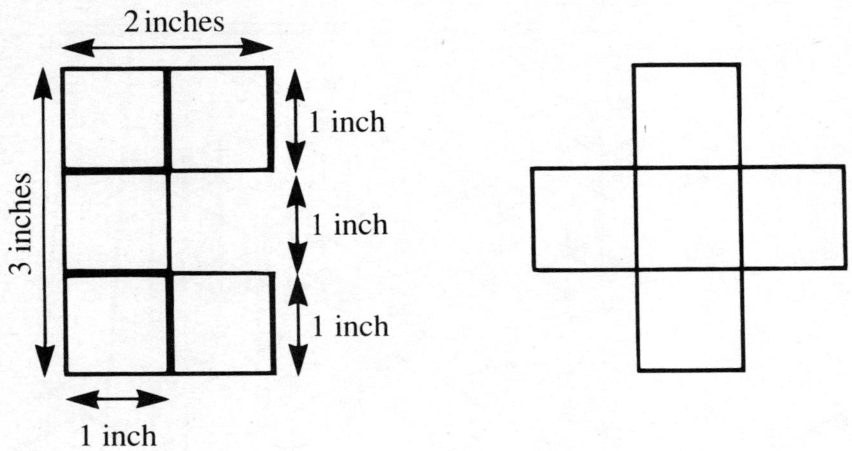

Ask your parent to help you draw pentominoes.

Pentominoes are a set of 12 shapes that are used in puzzles. Each shape is made up of 5 squares. Each pentomino follows these rules:

- The squares are the same size.
- Each square touches at least one other square.
- When squares touch, their sides match completely.

Cut out the pentominoes and see how many designs you can make by fitting various numbers of pentominoes together in different patterns. What designs can you make with 5 pentominoes? With 9? With 12? Use colors to make your designs more interesting.

If you want, you can save one of your designs. Just glue it on another sheet of construction paper. Or, if it is too big, tape the pieces together and turn the complete design over. Now you can hang your pentomino pattern on the wall!

Here are some other words that begin with *pent*: The Pentagon, pentathlon, pentangle, pentameter. Talk with your parent about what these words mean. Use your dictionary!

The Case of the Concealed Cubes

You need: colored pencils, crayons, or markers

Look at this pattern made with triangles and hexagons. Look very carefully—can you find the Concealed Cubes? Ask your parent to help you look until you can find them.

Now solve the Case of the Concealed Cubes so they won't ever get lost in the triangle/hexagon pattern again! Color the triangles and hexagons in each cube so that the cube shape stands out. If you want to, you may draw new lines in the pattern to help the cubes show up more clearly. How many cubes are there?

If you drew new lines in the pattern, what shapes did they make besides triangles and hexagons? Tell your parent what each new shape is called.

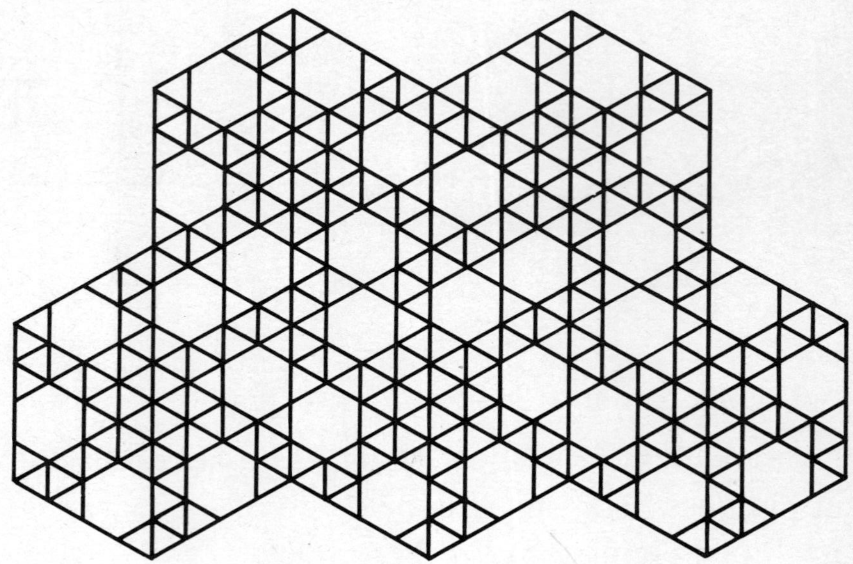

No Two Snowflakes Are the Same!

Follow these five steps to make a beautiful paper snowflake.

You need: one 3 × 5 index card, plain paper, construction paper, a ruler, a protractor, scissors, glue

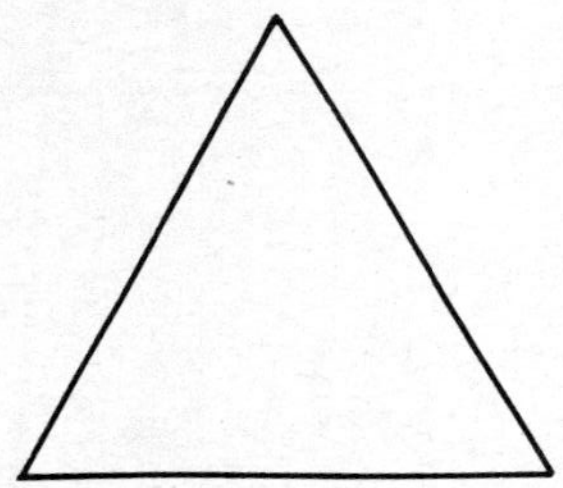

1. Make an equilateral triangle (2" × 2" × 2") on the index card; then cut it out:

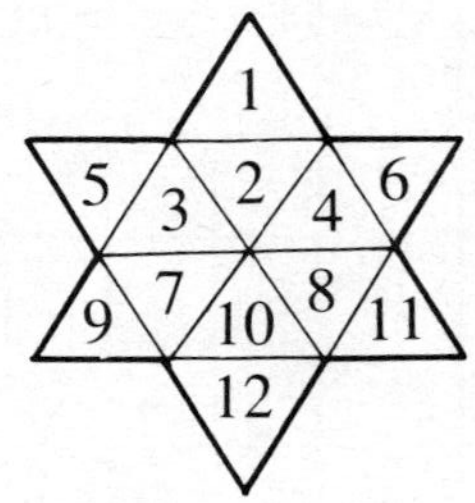

2. Trace around your triangle 12 times to make this pattern of a 6-pointed star:

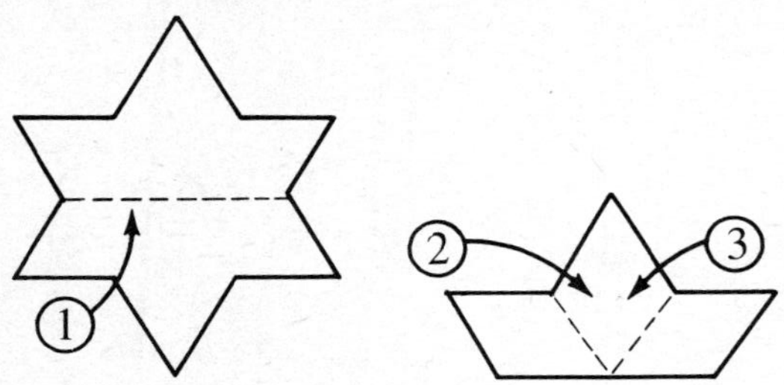

3. Cut out your 6-pointed star and fold it 3 times, like this:

4. Make small cuts in your folded star, like this:

5. Open your snowflake and glue it onto construction paper, like this:

Think about the lines of symmetry on your snowflake. How many are there?

Flipped Shapes

You need: a pencil, a small mirror

1. 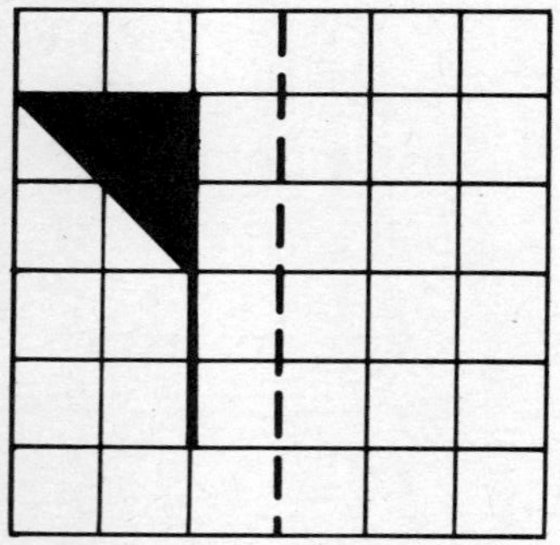

2. 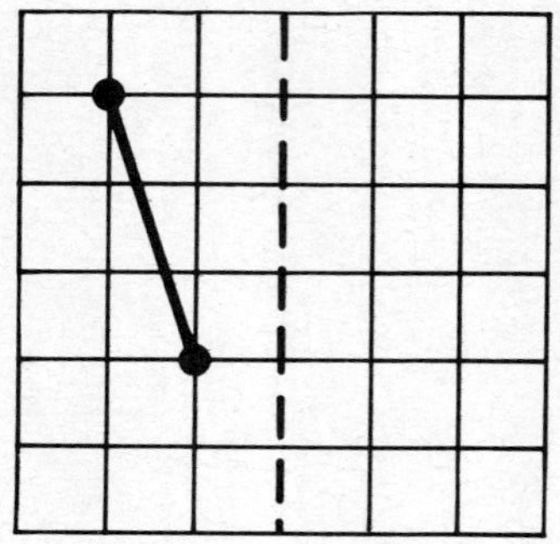

3. 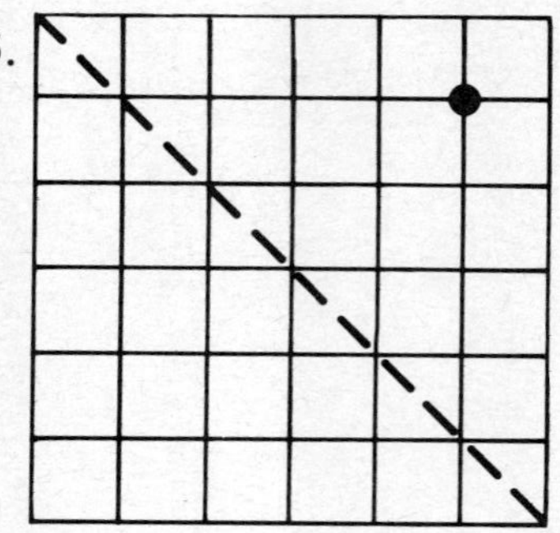

4. 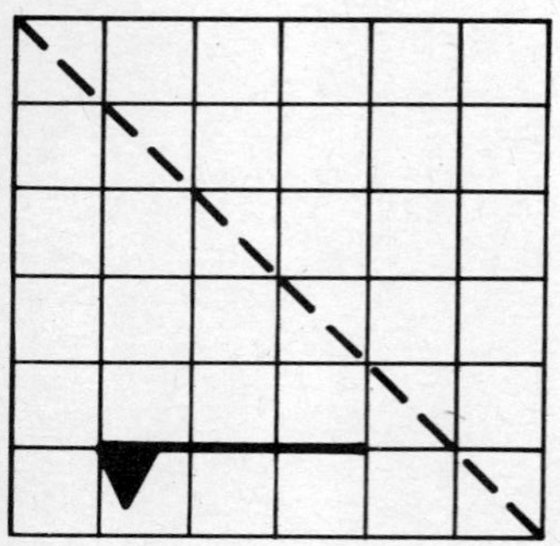

5. 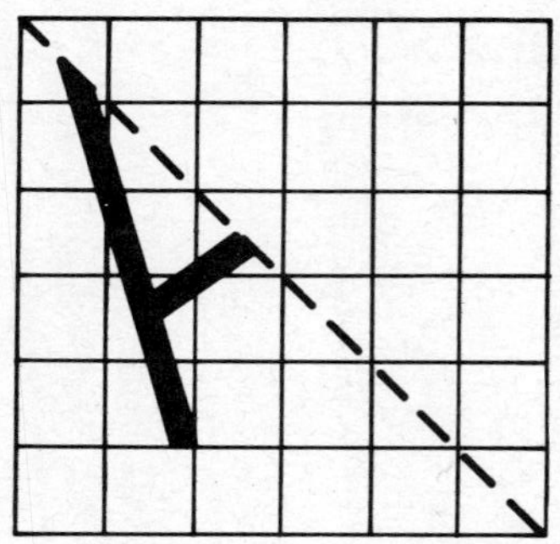

6. 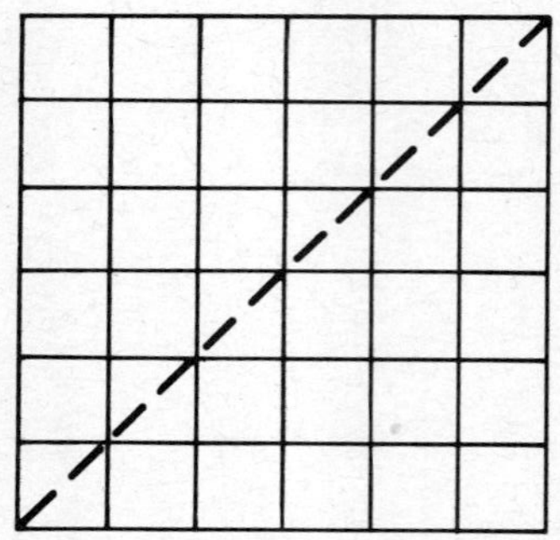

- Look at the shapes in Boxes 1–5. For each one, draw the shape's reflection on the other side of the dashed line.
- Put a small mirror on the dashed line to check your drawing. Does it match the mirror reflection? Fix your drawings that aren't "flipped" into a symmetrical pattern.
- Show your drawings to your parent. Draw a shape in Box 6, then ask him or her to draw its reflection.

PUZZLED??

Take turns with your parent filling in the puzzle.
Follow these rules:

- If you can fill in a word, score *one point*.
- If you can give a definition, score *2 points*, in total.
- If you can show an example (the top of a table could be a rectangle; for diameter, show the distance across a drinking glass), score *3 points* altogether.

Keep score below. Flip to see who goes first.

SCORE CARD	
Me	*My Parent*

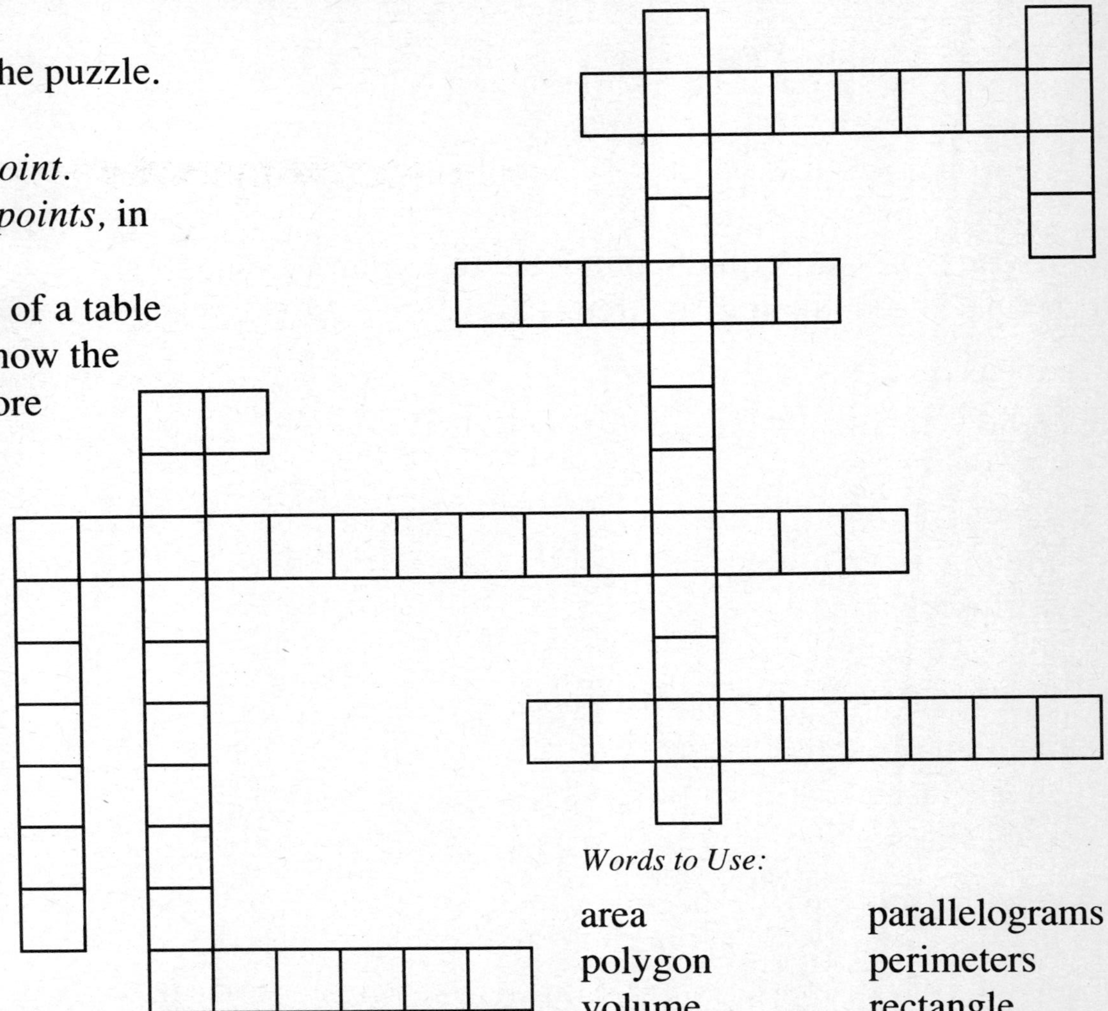

Words to Use:

area	parallelograms
polygon	perimeters
volume	rectangle
circumference	pi
diameter	square

Extended Activity

How Tall Is That Tree?

How would you measure the height of a tree? Would you use a tree-long ruler? Would you climb the tree with a measuring tape and count the yards as you go higher? Would you give a rope to a chimpanzee and ask him to climb the tree—then measure the rope when he reaches the top? Actually, you don't need to climb the tree or make friends with a monkey, and nobody makes a tree-long ruler!

To measure the height of a tree, you need a simple tool called a clinometer. Here's what you need to make a clinometer:

- a protractor
- a ruler
- a drinking straw
- adhesive tape
- a piece of cardboard or posterboard
- one foot of yarn or string
- metal washers or other weights that can be tied to the yarn or string

You will also need a yardstick or measuring tape to do this activity.

(How Tall Is That Tree? page 2)

How to Make a Clinometer

- On the cardboard or posterboard, draw a square with 6″ sides.
- Draw a diagonal line across the square to make a right triangle.

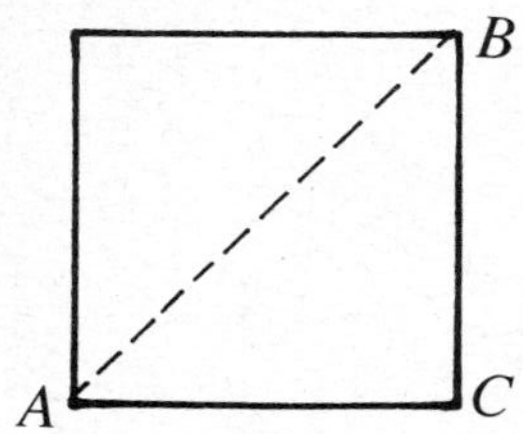

- Cut out the triangle (*ABC*).

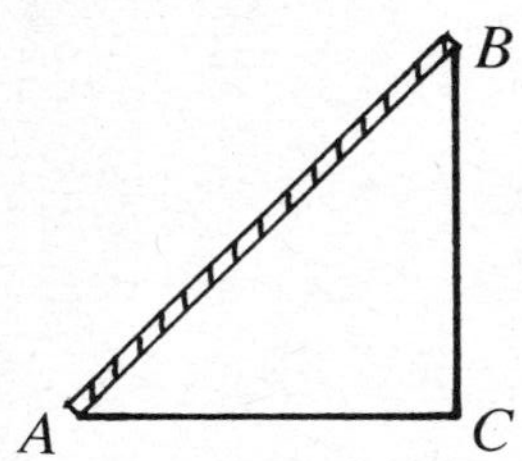

- Tape the drinking straw along *AB*.
- Make a hole $\frac{3}{4}$″ down from point *B* and $\frac{1}{4}$″ in from line *BC*.

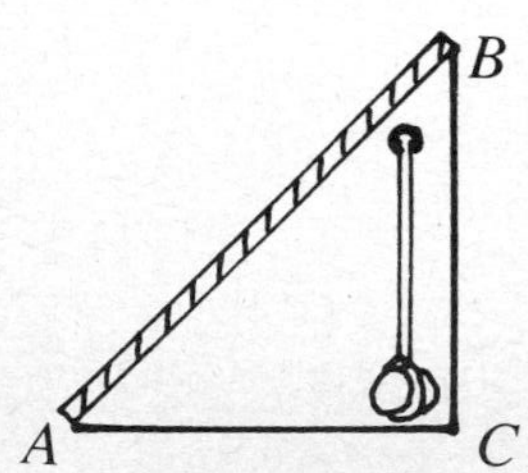

- Pull the yarn or string through the hole, then tie weights to both ends. Let them hang evenly on both sides.

How to Use a Clinometer

- Ask your parent to help you find a tall tree that is growing on flat land.
- Hold your clinometer so your yarn or string *always* stays parallel with edge *BC*. (Ask your parent to help check this.)

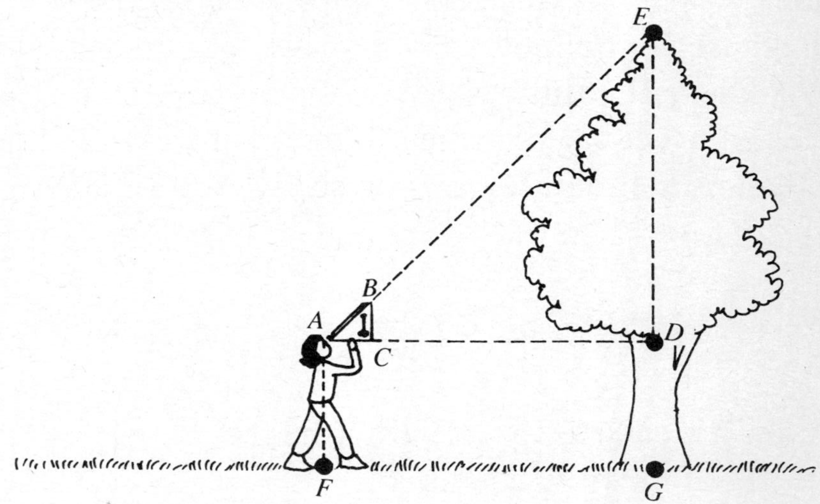

- Move to a position where you can see the top of the tree through the straw while keeping the yarn parallel to side *BC*.
- Ask your parent to measure either *AD* or *FG*, whichever is easier. Then ask your parent to measure *AF*. Write the measurements on the chart on the next page.

(How Tall Is That Tree? page 3)

Calculate Your Answer Here

Let's figure out how tall your tree is!

How long is *AD* (or *FG*)? __________

Because your clinometer was made as a right triangle, *ED* (which is part of the height of the tree) must equal your measurement of *AD* (or *FG*, which is the same). Can you see why *ED* equals *AD*?

How long is *AF*? __________

Now, it's easy to see that *DG* (the rest of the height of the tree) equals *AF*.

So, finally, the height of your tree is *ED* + *DG*.

How tall is your tree? __________

Draw Your Tree Here

Part 4

Fractions and Mixed Numbers: relating the parts to the whole

A note to parents

Real-life practice with fractions can be easy and delicious for you and your child. Just use a pizza! Your child may be surprised to learn that a slice of pizza is really a fraction in disguise! If pizza isn't a favorite at your house, you can substitute melons, apples, casserole dishes, or anything else that has a regular shape you can divide evenly.

Along with dividing a whole "something" into fractional parts, your child will also practice adding and subtracting those parts in Part 4.

Help your child build a strong fundamental understanding of fractions:

- Help your child use measuring cups and spoons to measure out various amounts of sugar or flour. Help him or her discover equivalent measures (for instance, 4 teaspoons is the equivalent of 8 half-teaspoons).
- Find a simple recipe, and help your child adjust the quantities of each ingredient to make twice or three times the number of servings.
- Point out to your child that a (conventional, not digital) clock is like a pizza with sixty slices, one slice for each minute. Then ask, "How many minutes are in one fourth of an hour?"
- Help your child practice changing whole numbers to mixed numbers. Ask him or her to change large numbers of minutes to hours and minutes. For example, ask, "How many hours and minutes is 162 minutes?" (2 hours, 42 minutes)
- Measure your child's height in inches. Then have him or her convert to feet and inches, a mixed number.

Spend just a fraction of your free time with your child on these and the following activities!

Gulp!

As your parent watches, make a mark 12 cm from the base of the container.

Fill your glass up to the mark with the liquid. Then, for this experiment:

- Calculate how many centimeters you will have to drink to leave $\frac{2}{3}$ of your drink still in the glass.
- Make a mark at this level. (Don't worry—it will wipe right off.)
- DRINK!
- Calculate how many centimeters you will have to drink to leave $\frac{1}{2}$ of what you started with.
- Make a mark at this level.
- DRINK!
- Without measuring, drink enough beverage to leave $\frac{1}{12}$ of your original drink in the container.
- Calculate how many centimeters make up $\frac{1}{12}$ of the original liquid.
- Measure the contents in the container.

How near were you in your estimate?

You need: 1 straight-sided glass at least 12 cm tall, water or your favorite beverage to drink, a colored marker

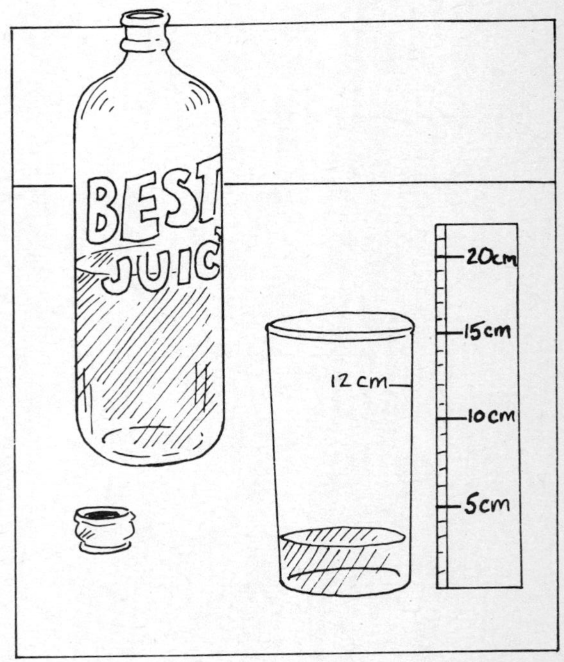

Why do you think you were asked to use a straight-sided container?

House Plan

You need: pencil and paper

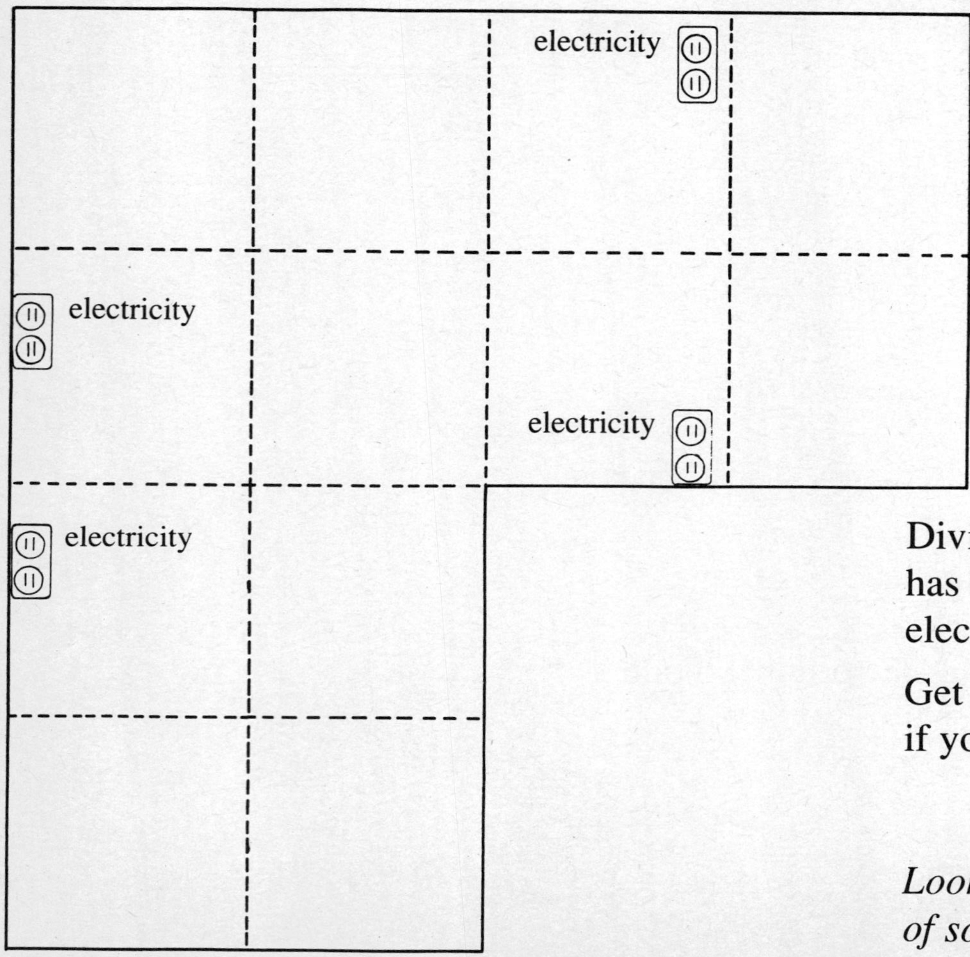

Divide this house plan into 4 rooms. Each room has to be equal in size and have its own supply of electricity. (They won't be square rooms.)

Get help from your parent in solving this problem, if you get stuck.

Look at the plan again. Can you find another way of solving the problem?

Make a Cat

Ask your parent to help you draw and cut out these 8 shapes. (You will need to measure the diameter of the circle carefully in order to know exactly what size to draw the square.)

Now arrange your shapes to make a cat!

You need: something circular with a diameter of from 1–3 inches, paper, a ruler, and scissors

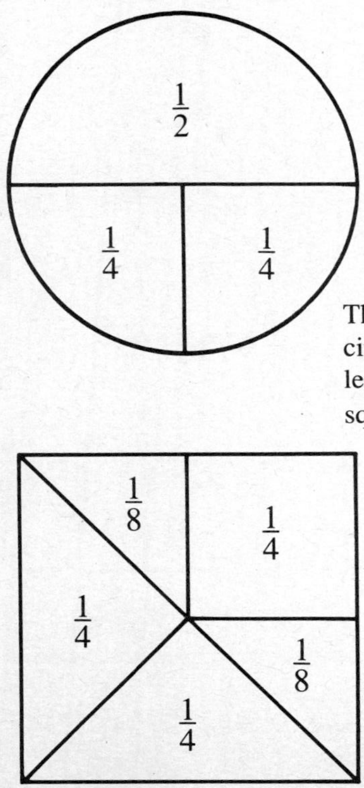

The diameter of the circle must be the same length as the side of the square.

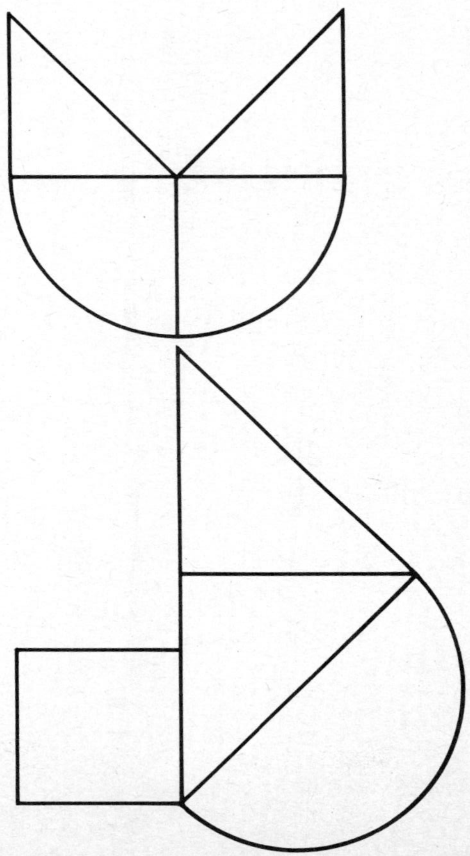

Why was it essential for the diameter of the circle and the sides of the square to be exactly the same length?

BIG IS LITTLE!

You need: 2 colored markers, a centimeter ruler

Work with your parent on this one.

- Find the 1-Line at left (the Whole Line).
- Measure in centimeters.
- Using your ruler as a guide, draw a colored line over the entire line.
- Notice the star at the top.
- Using another color of marker, draw a curved line joining all the stars from the top of the Whole Line, down to the $\frac{1}{20}$ line.
- What is the relationship between the length of each line and the denominator of the fraction?

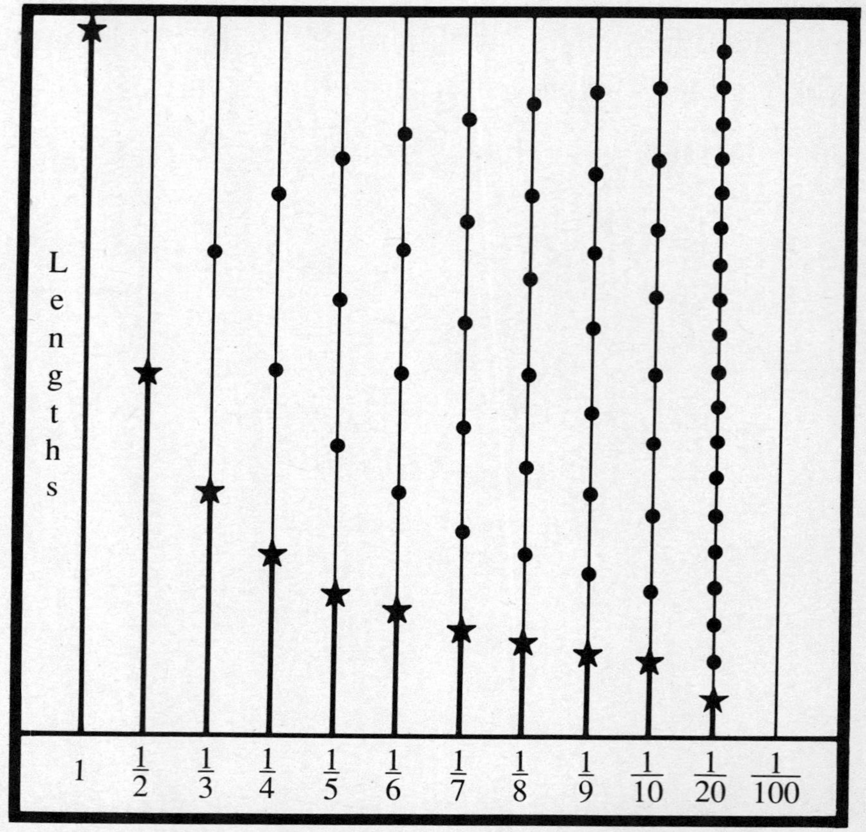

Using the curved line as a guide, draw a star where $\frac{1}{100}$ would be.

Snap!

Ask your parent to help you cut the index cards in half and mark each half card with one of the following numbers:

1	1	1	1	1	$\frac{1}{1}$	$\frac{2}{2}$	$\frac{3}{3}$
$\frac{4}{4}$	$\frac{5}{5}$	$\frac{1}{3}$	$\frac{2}{6}$	$\frac{3}{9}$	$\frac{4}{12}$	$\frac{5}{15}$	$\frac{2}{3}$
$\frac{4}{6}$	$\frac{6}{9}$	$\frac{8}{12}$	$\frac{10}{15}$	$\frac{3}{4}$	$\frac{6}{8}$	$\frac{9}{12}$	$\frac{12}{16}$
$\frac{15}{20}$	$\frac{75}{100}$	$\frac{1}{5}$	$\frac{2}{10}$	$\frac{3}{15}$	$\frac{4}{20}$	$\frac{10}{25}$	$\frac{40}{100}$
$\frac{3}{5}$	$\frac{6}{10}$	$\frac{9}{15}$	$\frac{12}{20}$	$\frac{15}{25}$	$\frac{60}{100}$	$\frac{1}{2}$	$\frac{2}{4}$
$\frac{3}{6}$	$\frac{4}{8}$	$\frac{5}{10}$	$\frac{50}{100}$	$\frac{1}{4}$	$\frac{2}{8}$	$\frac{3}{12}$	$\frac{4}{16}$
$\frac{5}{20}$	$\frac{25}{100}$	$\frac{5}{25}$	$\frac{20}{100}$	$\frac{2}{5}$	$\frac{4}{10}$	$\frac{6}{15}$	$\frac{8}{20}$

You need: 28 index cards, scissors, a marker

Shuffle the cards thoroughly.

Deal them out between your parent and yourself.

Then, to play:

- Take turns placing a card face up on the table.
- Lay out the cards side by side.
- When you see an equivalence, shout "SNAP!" The first person to shout collects all the cards on the table at that point.
- Start turning up cards again.

The winner is the person with more cards.

What can you do to both the numerator and denominator of a fraction to make another equivalent fraction?

THROW A HALF

NUMBER BOARD

2	4
6	8
10	12

You need: a number cube

Here's how to play the game:

- Roll the cube onto the number board, to make a fraction.
- Make the number on the cube the numerator (top number).
- Make the board number the denominator (bottom number). You may make the board number the numerator if an improper fraction occurs. (If the cube falls on a line, you may choose any square touched by the cube.)
- Take turns with your parent to have 10 throws each. Write down each fraction you make.

The player who throws more halves is the winner. (Of course, fractions equivalent to $\frac{1}{2}$ count.)

Is this a game of mostly skill or luck? Why?

Twice as Much

Read the list of ingredients for Honey Flapjacks to your parent.

Imagine you are about to make these flapjacks, except you need to make enough for 2 platefuls, not just one.

Enter in the boxes the new amounts of each ingredient you would need.

Honey Flapjacks

$\frac{1}{2}$ c ☐ margarine
$\frac{2}{3}$ c ☐ light brown sugar
$\frac{3}{4}$ c ☐ honey
$3\frac{1}{2}$ c ☐ rolled oats
$1\frac{1}{4}$ c ☐ milk

Pentagons

You need: a centimeter ruler, a pencil, a strip of paper 11" × 1"

How to Make a Pentagon

Make a knot in a strip of paper.

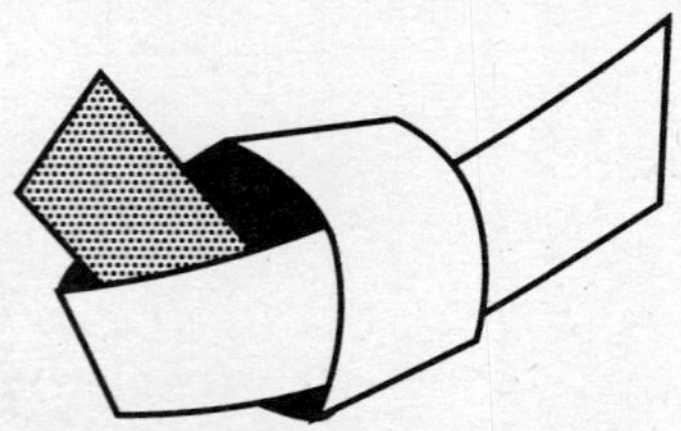

Flatten the knot.

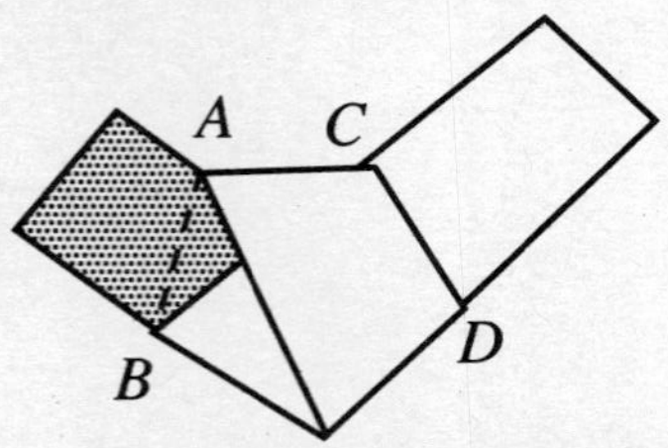

Cut along lines *AB*, *CD* to finish the pentagon.

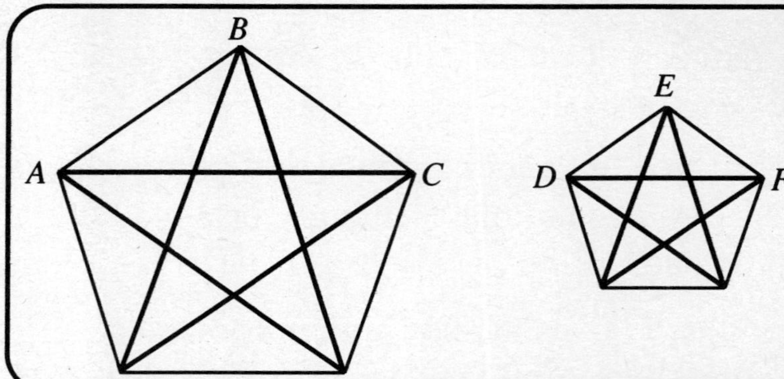

Ask your parent to help you measure the line segments *AC*, *BC*, *DF*, and *EF* in centimeters.

Enter your measurements here:

AC→ ☐ : ☐ ←*BC*

DF→ ☐ : ☐ ←*EF*

What do you notice about the two ratios? Discuss your conclusion with your parent.

What are you wearing today?

Take a look at what you and your parent are wearing today.

Decide which items of clothing are made from fabric, which from leather, and which from both fabric and leather.

Write the names of the articles on this Venn diagram.

You need: a marker, paper and pencil

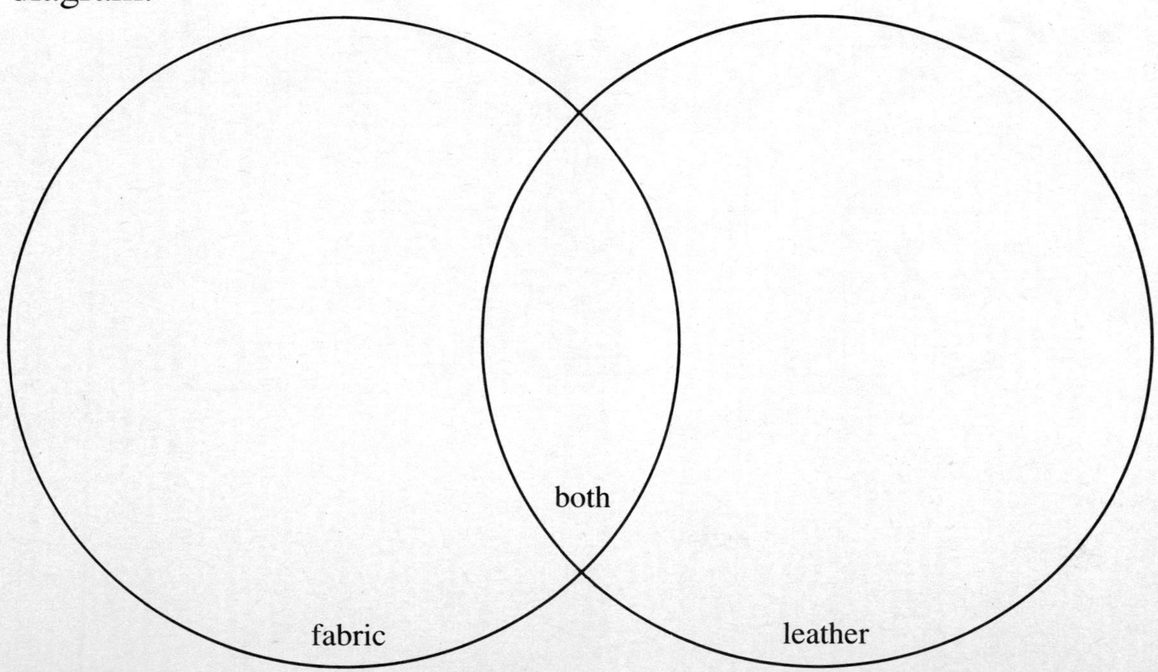

Where would you draw items of clothing which are neither fabric nor leather?

Extended Activity

Scavenge for Fractions

Go on a scavenger hunt with your parent, either in a do-it-yourself (home improvement) store or a fabric store.

Take the appropriate checklist from the next page with you. Use the checklist to record the fractions you find.

FRACTIONS I HUNTED DOWN (Scavenge for Fractions, page 2)

Beside each item you find, make a check mark. Write any fractions you see on the label, the bin, or wherever. You may not find every item on the list. In the blank spaces at the bottom of the list, add two items that you found on your own.

A Fabric Store			
No.	*Item Name*	✔	*Fractions You Have Seen*
1	tape (hem tape)		
2	buttons		
3	snap fasteners		
4	pins		
5	elastic		
6	yarn		
7	cotton piping		
8	tacks		
9	ribbon		
10	yardstick		
11			
12			

Why are exact measurements important in a fabric store?

A Do-It-Yourself Store			
No.	*Item Name*	✔	*Fractions You Have Seen*
1	furniture		
2	stainless steel sink		
3	lighting fixture		
4	rubber mat		
5	conduit piping		
6	scotch tape		
7	bug fighter		
8	screws		
9	nylon rope		
10	insulation tape		
11			
12			

Why do you see such small fractions in a do-it-yourself store?

A HUNTING STORY (Scavenge for Fractions, page 3)

On this page, write a summary of your fraction scavenger hunt. Tell where you went, what you found, what you learned. Did you find any uses of fractions that were new to you? What did you learn from your parent about fractions and their uses? Include all this information in your summary, if you can fit it in.

Straight Becomes Curved

You need: one index card cut to 3″ × 3″, a piece of plain paper, a black marker, a ruler, scissors, glue

To do this trick:

- On the 3-inch-square card, draw a series of diagonal lines.
- Completely blacken every other diagonal strip and draw different patterns on the strips in between.
- Rule vertical lines on the drawing, $\frac{1}{4}$ inch apart.
- Carefully cut along the vertical lines and, as you do, place them in order on the piece of paper.
- Arrange the strips into a curved pattern; then glue them down.
- Show your parent and describe how you made it.

What fraction of the whole of your pattern is one strip?

Write an addition or a subtraction fact about your pattern like the one in the example.

Use the same method to experiment with magazine pictures.

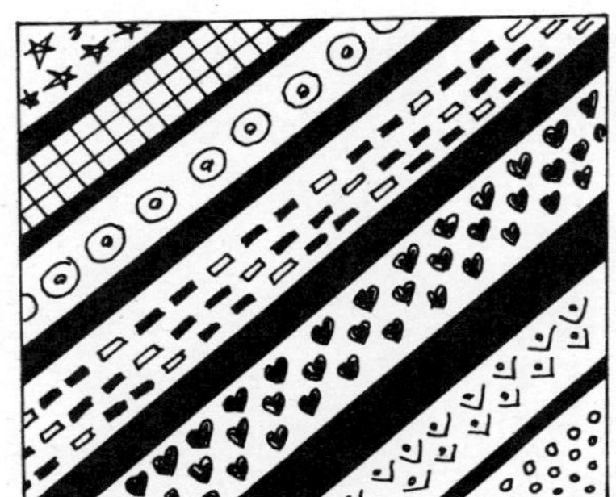

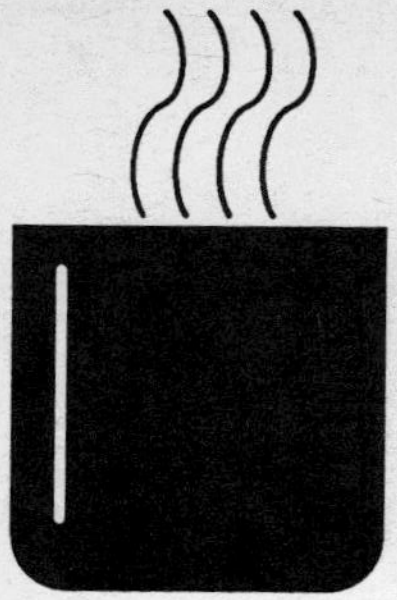

We're Cooking With Fractions

In this activity, you and your parent can make two tasty seasonings for family meals and study fractions at the same time!

- If you don't have all the ingredients, you can either leave something out or ask your parent what would be a good substitute.
- You can also just use salt in this activity. Then when you are finished, you can put it back in its container.

You need: 2 medium-sized bowls, a set of measuring spoons, 2 teaspoons, 2 small jars with lids, the ingredients in the recipes below

RECIPE #1: For spicy vegetable and pasta dishes

3 teaspoons	dried parsley flakes
$2\frac{1}{4}$ teaspoons	dried chives
4 teaspoons	grated parmesan cheese
$\frac{3}{4}$ teaspoon	garlic powder or minced garlic
$1\frac{1}{4}$ teaspoons	dry mustard
$\frac{1}{4}$ teaspoon	salt
$\frac{3}{4}$ teaspoon	black pepper

RECIPE #2: For sweet toast, rice, or cereal

7 teaspoons	brown sugar
2 teaspoons	cocoa powder
$1\frac{1}{4}$ teaspoons	cinnamon powder
$\frac{1}{4}$ teaspoon	nutmeg
$\frac{1}{8}$ teaspoon	clove powder

Note: 3 teaspoons = 1 tablespoon

Add and mix the ingredients for each recipe in separate bowls.

Find the total measurement of the ingredients in each recipe, then measure the contents of each bowl as you put it in a jar.

Was your addition correct? Store the jars in your refrigerator.

RECIPE #1 Total __________

RECIPE #2 Total __________

JUMP-LINE, JUMP!

You need: a newspaper, a ruler, pencil and paper

Look at the front page of a newspaper with your parent.

Find a story with a jumpline (a jumpline tells you where the story continues). Then, find the continued story.

Measure the length of the beginning of the story on the front page and the ending on the continued page, in inches. Write your measurements below.

- Length of story on front page = ☐ inches
- Length of story on continued page = ☐ inches
- Total length of story = ☐ inches

Now, figure the fractions and write them below.

- Fraction of story on front page = ☐
- Fraction of story on continued page = ☐

Why do editors print only a fraction of a story on the front page?

Musical Fractions

g1

1 ¦ $\frac{1}{4} + \frac{1}{16} + \frac{1}{16} + \frac{1}{16} + \frac{1}{16} + \frac{1}{8} + \frac{1}{8} + \frac{1}{8} + \frac{1}{8} = 1$ ¦ $\frac{1}{2} + \frac{1}{4} + \frac{1}{8} + \frac{1}{8} = 1$ ¦ 1

Look at this line of music with your parent. Can you see the three vertical lines that mark off the musical measures?

The value for each measure must equal 1. Check that each measure above does, by adding the fractions below the notes.

g2

- On this line of music, give each note its numerical value. Use the legend to help you.
- Now, add the fractions to discover where to put in the measure lines on this score. Every time the note values add to 1, draw a vertical line. How many measures are there?

Legend

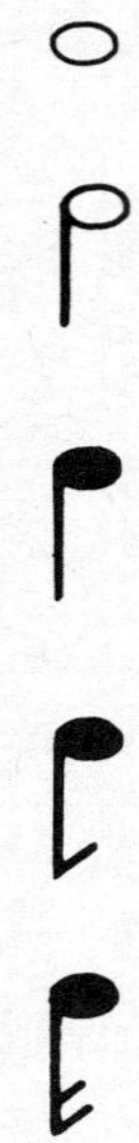

= 1 whole measure

= $\frac{1}{2}$ measure

= $\frac{1}{4}$ measure

= $\frac{1}{8}$ measure

= $\frac{1}{16}$ measure

Look at a piece of written music in a store, at the library, at school, or at home. See if you can recognize the notes.

Magic Matching Fractions!

Magic squares are magic because they add up to the same total, no matter which direction you calculate—across, down, or diagonally!

You and your parent play this game to make your own magic squares.

- Choose one of the unfinished squares below to play on. Your parent gets the other one.
- One of you say "GO!" to start the game.
- Fill in all the boxes with the correct fractions as fast as you can!

What is the total for your magic square? For your parent's?

Magic Square #1

$\frac{1}{4}$		$\frac{1}{8}$	
	$\frac{5}{16}$		
$\frac{1}{2}$		$\frac{3}{8}$	

Magic Square #2

$\frac{2}{3}$		$\frac{1}{3}$	
	$\frac{5}{12}$		
$\frac{1}{2}$		$\frac{1}{6}$	

What denominator was the key to your magic square? What was the key to your parent's magic square? Can you make a new magic square with the key denominator of 8?

Puzzle Pictures

Test your problem-solving skills with this fraction puzzle!

Look at the puzzle below with your parent. Use the clues to find the fraction that each shape represents.

Remember to write each fraction in its simplest form.

Puzzle

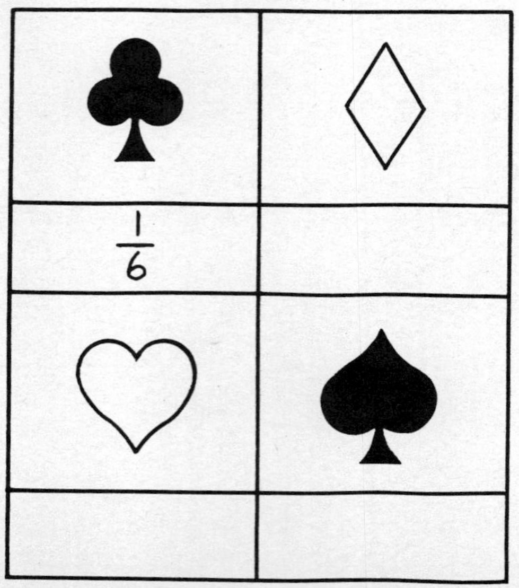

Clues

Make up your own fraction puzzle and ask your parent to find the answers.

BREAK THE CODE!

Use the clues below and the Code Breaker to discover the secret message.

The Secret Message

1	2	3	4
5	6		
7	8	9	

The Code Breaker

A	B	C	D	E	F	G	H	I	J	K	L	M
1	2	3	4	5	6	7	8	9	10	11	12	13
N	O	P	Q	R	S	T	U	V	W	X	Y	Z
14	15	16	17	18	19	20	21	22	23	24	25	26

Find the nearest whole number for your clues.

1. $12\frac{7}{8}$
2. $\frac{7}{8}$
3. $19\frac{15}{16}$
4. $7\frac{4}{5}$
5. $8\frac{5}{6}$
6. $19\frac{3}{7}$
7. $6\frac{1}{9}$
8. $20\frac{6}{10}$
9. $14\frac{4}{9}$

Invent your own message now, and write the clues as mixed fractions, which have to be rounded to the nearest whole number. Ask your parent to decode your message.

Angles in Time

You need: a protractor

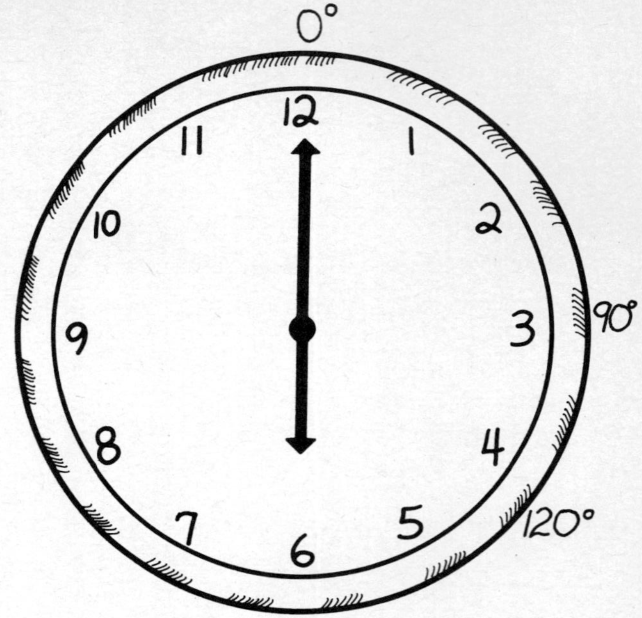

Look at this clock with your parent. How many equal parts does it have?

Draw lines out from the center to divide the clock face into parts. Then measure the angle of each division.

- How many degrees in one division? ________
- How many degrees total in all divisions? _____

Imagine living in a place where people tell time by *degrees* instead of by hours and minutes! Can you translate what these people tell you?

Hours

- Let's meet for lunch at 0° ________
- School begins at 270° ________ and ends at 90° ________
- My favorite TV show comes on at 225° ________

Minutes

- My bus will come in 60° ________
- I'm so hungry, and dinner is still 120° away! ________
- The speech was long—300°! ________
- May I have 30° of your time? ________

Between the times of 1:25 p.m. and 1:40 p.m., how many degrees does a minute hand travel?

Make My Hexagon Simple

You and your parent can make a hexagonal spinner in this activity and then use it to play a game in the next activity!

You need: a 3 × 5 index card, plain paper, ruler, compass, protractor, scissors, glue, pencil, and a paper clip

1. On paper, draw a 3″ diameter circle. Cut it out and fold it into quarters. Mark the center point.

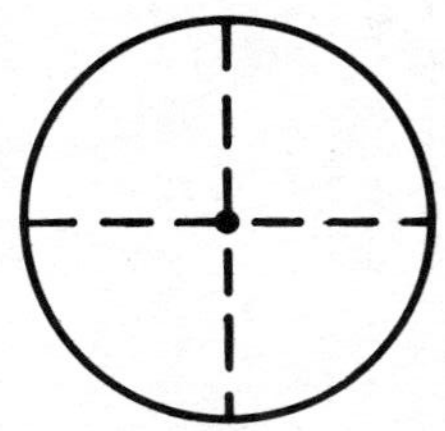

2. Fold a section of the circle to touch the center point.

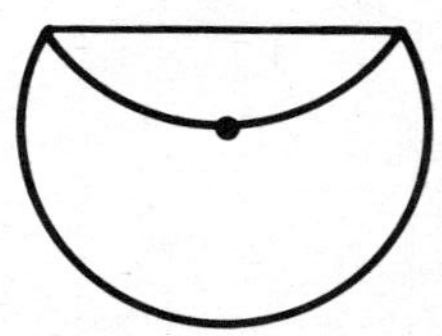

3. Fold twice more until you have an equilateral triangle.

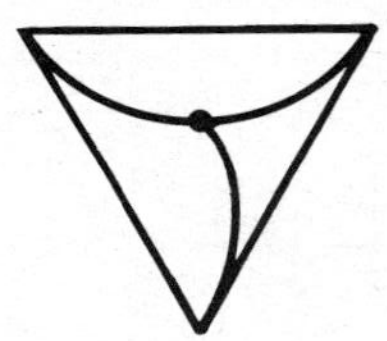

4. Unfold the circle and draw in the line segments *AD*, *BE*, and *CF*.

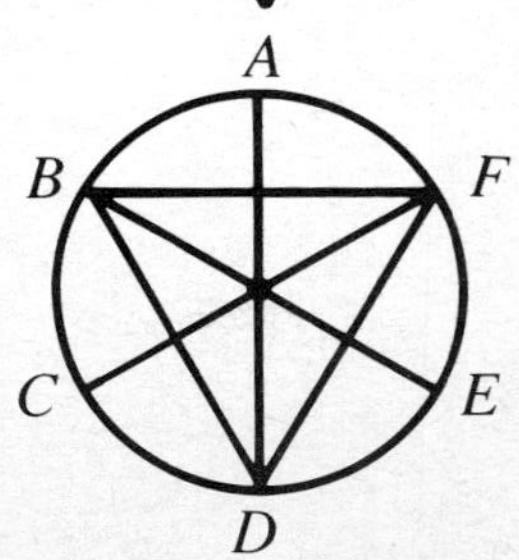

5. Draw segments *AB*, *BC*, *CD*, *DE*, *EF*, and *FA*. See the hexagon?

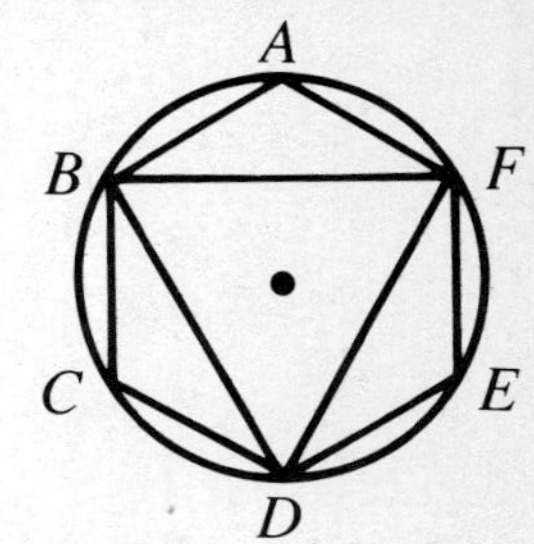

6. Glue your hexagon to a card, then cut it out. Write the numbers 1–6 on it.

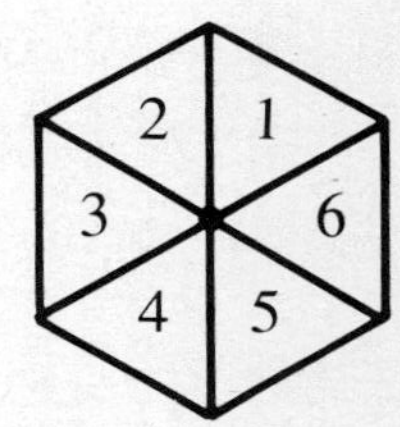

Measure the angles of your hexagonal spinner.

How many degrees does each of the six parts have?

My Fraction, Please!

You and your parent can play this game with your new spinner!

Take turns making fractions with the number cube and the spinner.

- Roll the number cube to find the numerator (top number).
- Spin the spinner to find the denominator (bottom number).

The first person to get 10 points is the winner!

You need: your new spinner (from page 143), a number cube, pencil and paper, and a paper clip

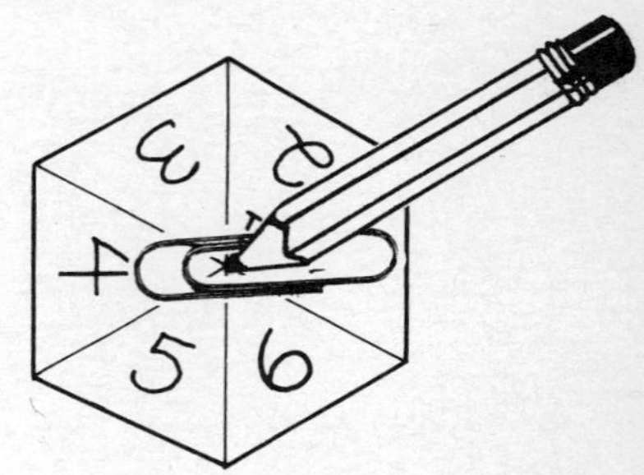

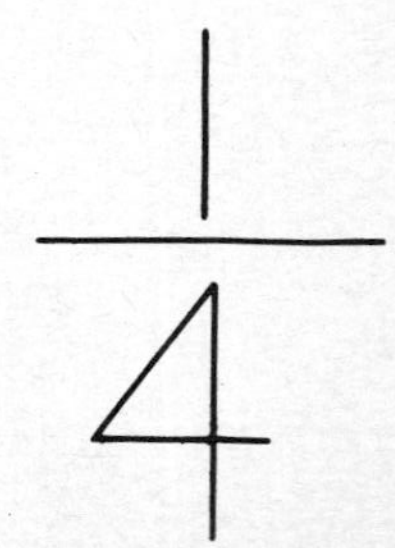

How to Score

- If the fraction is in its simplest form you get 2 points. Example: $\frac{3}{4}$ = 2 points to me
- If the fraction is *not* in its simplest form your parent gets 1 point. Example: $\frac{4}{6}$ = 1 point to parent
- If the fraction is improper then the player gets *no* points. Example: $\frac{6}{4}$ = 0 points

Extended Activity

Weigh Your Options... At the Post Office

Imagine you know this family very well. They are an American family living in West Germany. October has come—birthday month for Mr. Miller and Joe. It's also the month for Halloween and almost time for Thanksgiving. You have decided to mail birthday cards to Mr. Miller and Joe, Halloween cards to the three children, and a Thanksgiving card to the whole family.

You would prefer to mail each card separately, so it can arrive at the right time. But you know this may be too expensive. Sending all the cards together may be more economical. You need to visit the post office to help you decide which option to choose.

Collect together:

3 Halloween cards 1 Thanksgiving card
2 birthday cards
(Your cards can be new, used, or homemade.)
6 small envelopes 1 big envelope

Ask your parent to take you to the post office with your cards and envelopes.

IN THE SCALE (Weigh Your Options..., page 2)

Complete this data sheet at the post office.

- Locate the weighing machine for mail up to 1 pound.
- Discuss the divisions on the scales with your parent. Look for fractions of $\frac{1}{16}$ and $\frac{1}{32}$ of a pound.
- Weigh your cards and envelopes, and enter the data below:

Cards	*Weight*
Leroy's Halloween	
Maria's Halloween	
Joe's Halloween	
Father's birthday	
Joe's birthday	
Thanksgiving	
1 big envelope	
6 small envelopes	
Total Weight	

Remember: The post office requires you to round up to the next $\frac{1}{2}$ ounce.

MAKING THE DECISION (Weigh Your Options..., page 3)

If it costs 45¢ for each half ounce of airmail sent to Europe, calculate the two costs and enter them below. (Remember not to count the weight of the big envelope in the second calculation.)

It will cost ______ to mail all the cards in one envelope.

It will cost ______ to mail the cards separately.

Which is more expensive?

Draw a picture of the post office scale, complete with the marked divisions or the digital readout.

What fraction of a pound is $\frac{1}{2}$ ounce? What fraction of a pound is 1 ounce?

Part 5

Ratios, Averages, and Probability: numbers in the news

A note to parents

Your child is already quite familiar with ratios, though he or she may not know it. A ratio is simply a fraction. Slicing a birthday cake, dividing baseball cards among the gang, getting a saving of one fourth off the price of a new pair of jeans—these are all example of real-world ratios. Part 5 contains more practice with ratios, plus practice using averages and probabilities.

Try these simple activities to give your child more practice with ratios:

- Have your child find sales ads in a newspaper. Ask him or her to calculate the actual dollar saving involved for each.
- Have your child help you comparison shop on your next shopping trip. Your child will take pride in helping you find the best buys!

These activities will help your child make sense of averages and probabilities:

- Point out occasions in everyday life in which averages and probabilities are used. (Hint: weather forecasting is a good place to begin.)
- Listen to television or radio news reports together, and have your child write down any statistics you hear. Talk with your child about what the numbers actually mean.
- Find examples of statistics in ads. Discuss them together, and be sure to talk about why advertisers use statistics.

With these activities and those that follow, your child can successfully deal with numbers in the news!

HOW LOW CAN YOU GO?

You need: these colored pencils or crayons: red, orange, yellow, green, blue, violet

The figures at right represent fractions. There are at least two fractions of equal value (equivalent fractions) represented in each figure. In some cases there are three.

The first figure is done for you. Find the fractions in figures 2 through 7. (*Hint:* Make sure fractions you have found are reduced to lowest terms.) Ask your parent to check your work.

Using a different colored pencil for each of the remaining six figures, color in the white spaces after you find the fractions.

In what natural phenomenon do you see all these colors at once?

1

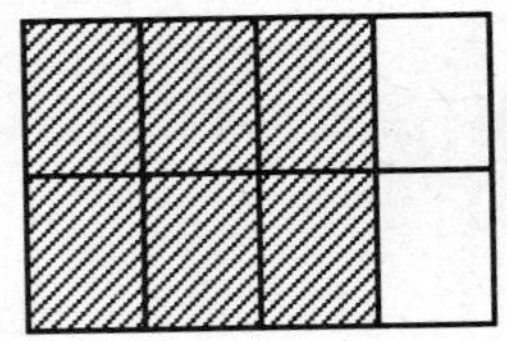

Example $\frac{6}{8}$ and $\frac{3}{4}$

2

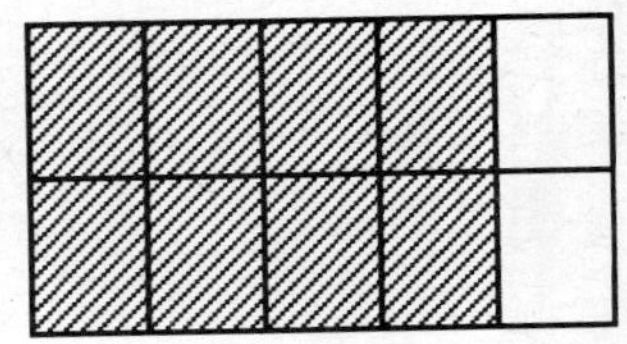

3

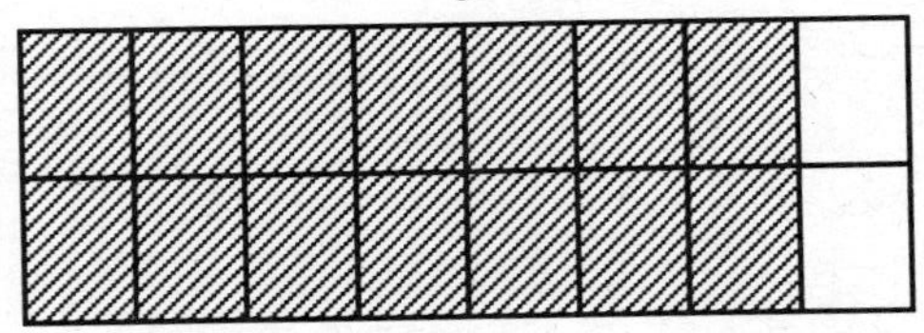

4

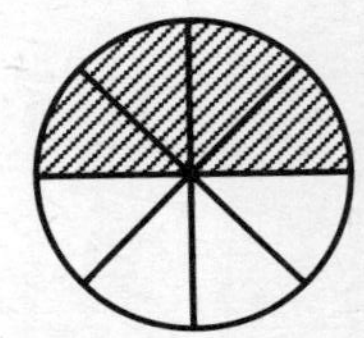

5

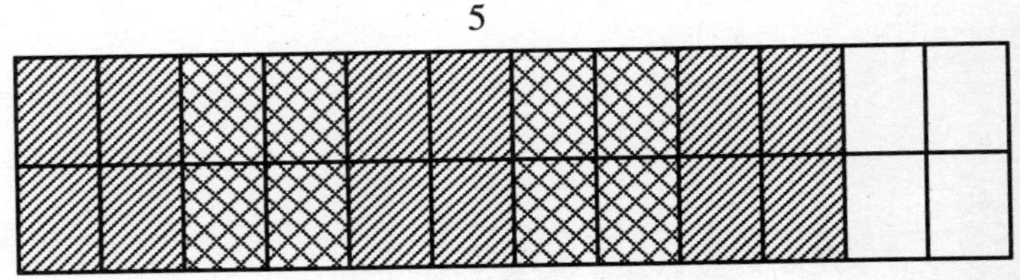

6

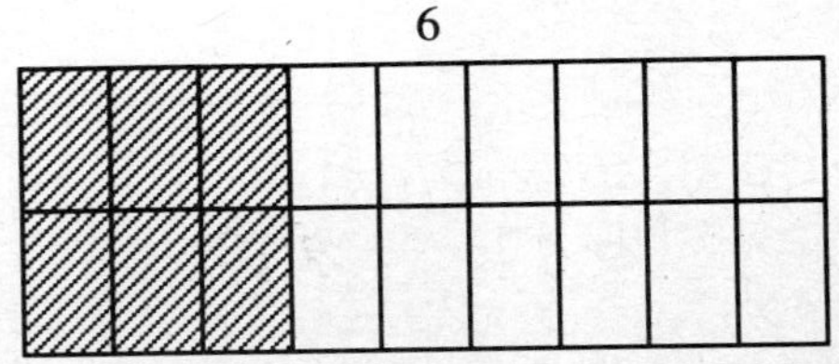

7

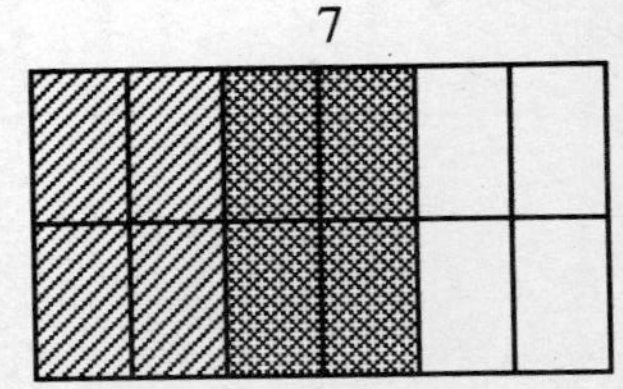

A SQUARE DEAL

This is a game of accuracy and speed. Each player selects a card below. At the "Go!" signal, both you and your parent begin multiplying the numbers from the vertical and horizontal rows. Write answers in the corresponding boxes. (For example, the product of $\frac{3}{8} \times 6$ goes in the starred box on Card A.) The first player to *correctly* solve all the problems is the winner.

A	$\frac{3}{8}$	$\frac{1}{4}$	$\frac{5}{6}$
6	★		
$\frac{2}{3}$			
$\frac{4}{5}$			

B	$\frac{3}{5}$	$\frac{2}{7}$	$\frac{5}{6}$
$\frac{3}{4}$			
4			
$\frac{1}{8}$			

If you were able to solve any of the problems on your card mentally, tell which ones and explain how.

Dividing by Birthdays

Make fractions out of all the birthdays. Use the number of the month as the numerator (top number) and the day as the denominator (bottom number). For example, April 11 becomes the fraction $\frac{4}{11}$.

To play:

- You and your parent each write as many different division problems as you can in one minute, using the birthdate fractions.
- Exchange papers and work the division problems.
- Check your answers.

You need: paper, pencil, and the birthdates of you and four of your friends or family members

Is April 11 ÷ September 4 the same as September 4 ÷ April 11?

SCALE IT DOWN

You and your parent *each* measure your kitchen and make a scale drawing. Here's how:

- Step off the distance of each wall of the kitchen. Take steps so that one foot touches the next one.
- Then draw your scale map of the kitchen on your piece of paper. Make 1 step = 2 cm or $\frac{1}{2}$ inch.
- Your parent now steps off the distance of each wall of the kitchen. He or she should take the same kind of steps.
- Your parent should make a scale drawing of the kitchen on the other sheet of paper. Make 1 step = 2 cm or $\frac{1}{2}$ inch, again.

You need: an inch or centimeter ruler, 2 sheets of paper, a pencil

Whose drawing is bigger? Discuss why.

The Daily Round

You need: a pencil, a protractor

Discuss with your parent what you have learned in school about circle graphs.

Remember each circle represents one whole hour, one whole day, or one whole of anything. Remember also that parts of the circle graph are fractions of 360°, which is a complete turn.

This is how Marco spends his day.

Activity	*Sleep*	*School*	*Travel*	*Eating*	*TV*	*Misc*
Hours	9	8	2	2	2	1
Degrees	135°					

One complete 24-hour day is represented by 360°.

Therefore, one hour is 15° (360° ÷ 24 = 15°).
If Marco sleeps 9 hours, the number of degrees used to represent 9 hours is 135° (9 × 15°).

135° has been drawn for you on the circle graph.

Ask your parent to help you calculate the number of degrees which represent each activity in Marco's day.

Then measure the correct number of degrees with a protractor, and write the name of the activity on the circle graph.

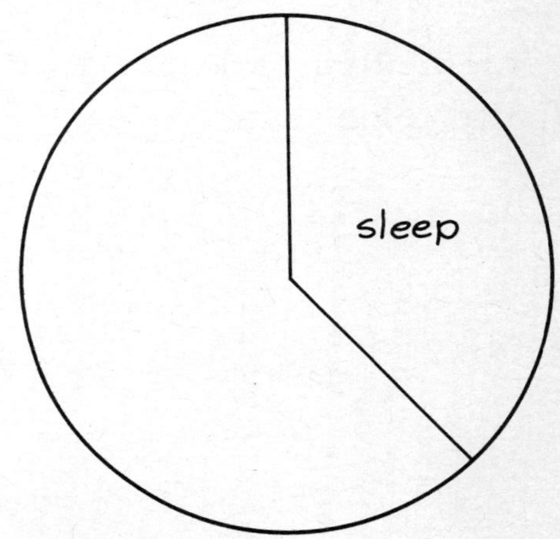

Sometimes circle graphs are called pie charts. Why do you think this is so?

Extended Activity

Where Does My Money Go?

You need: a pencil, colored pencils, a protractor

Imagine it's your birthday, and you have been given a total of $60 from various members of the family.

Discuss with your parent how much you would save and/or spend of that much money.

In the boxes below, tell the method you would use to save some of the money and draw the items you would buy.

Enter the amount of money involved. Use catalogs and your visits to stores to help you be accurate in writing your costs.

Amount saved:	*Amount spent:*	*Amount spent:*	*Amount spent:*

On the next two pages you will practice a good way to show where your money goes.

GETTING READY TO SHOW . . . (Where Does My Money Go? page 2)

Ask your parent to help you with the calculations you need to make to draw the circle graph on the next page. (A calculator would be helpful but is not essential for the calculations.)

	Example				
Name of item	*Sport timer*				
Amount of money spent or saved	\$12.00				
Fraction of \$60	$\frac{12}{60} = 0.2$				
Degrees which would be needed to represent the fraction on the circle graph	$0.2 \times 360 = 72°$				

HERE IT IS (Where Does My Money Go? page 3)

With a protractor, measure the necessary degrees to represent the part of $60 that each amount you saved or spent stands for.

Label each section.

What do you think the advantages of circle graphs are?

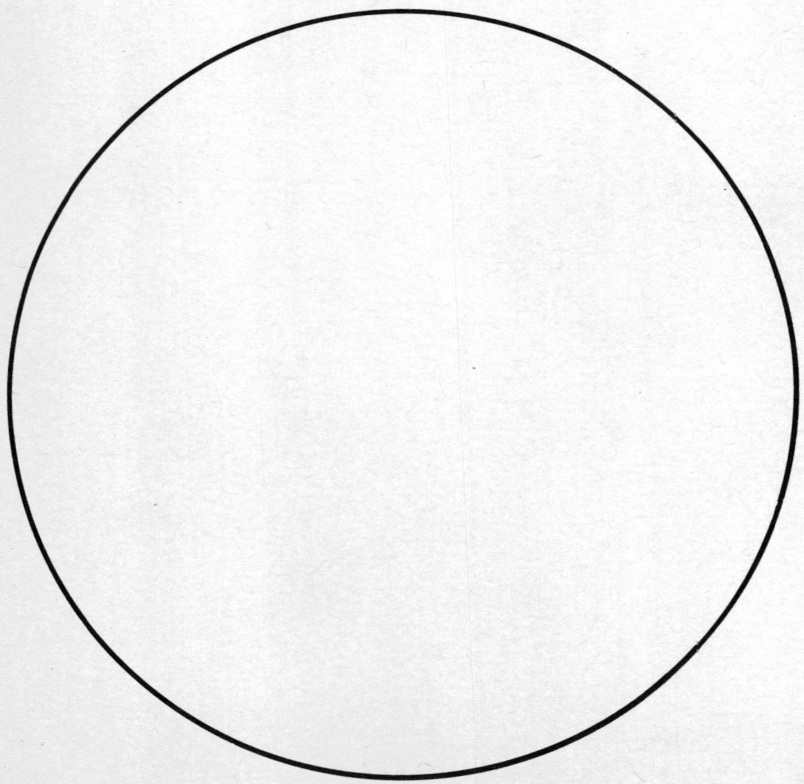

Scavenger Hunt Ratios

Look around your house and find these ratios.

1. What is the ratio of doors to windows?

   ☐ doors
   ―――――
   ☐ windows

2. What is the ratio of tables to chairs?

3. What is the ratio of cans to boxes in your kitchen?

Think of two other ratios of items in your house. Write them here.

a. ☐ ________
   ―――――
   ☐ ________

b. ☐ ________
   ―――――
   ☐ ________

Speaking of Ratios

Example:

CHRISTY HAYWOOD

Vowels are *I, A, O, O, Y, Y*
Consonants are *C, H, R, S, T, H, W, D*
Total letters = 14
Ratio of vowels to consonants = $\frac{6}{8}$

Write the names of three of your family members

1.

2.

3.

Discuss with your parent:

- What is the ratio of vowels to consonants in each name?
- What is the ratio of vowels to all letters in each name?
- What is the ratio of consonants to all letters in each name?

Can you reduce any of your ratios?

Different, and Yet the Same

You need: pencil and paper, a centimeter ruler

Look at this right-angled triangle with your parent.

How many triangles can you see?

- Look for the ratio of the vertical to the horizontal measurements, for each of the 3 triangles.
- Make a note of the ratios and compare them.
- Draw 3 new horizontal lines on triangle *AFG*. Draw them anywhere.
- Measure the vertical and horizontal sides of the new triangles you created. Record them.
- How do the new ratios compare with the old ones?

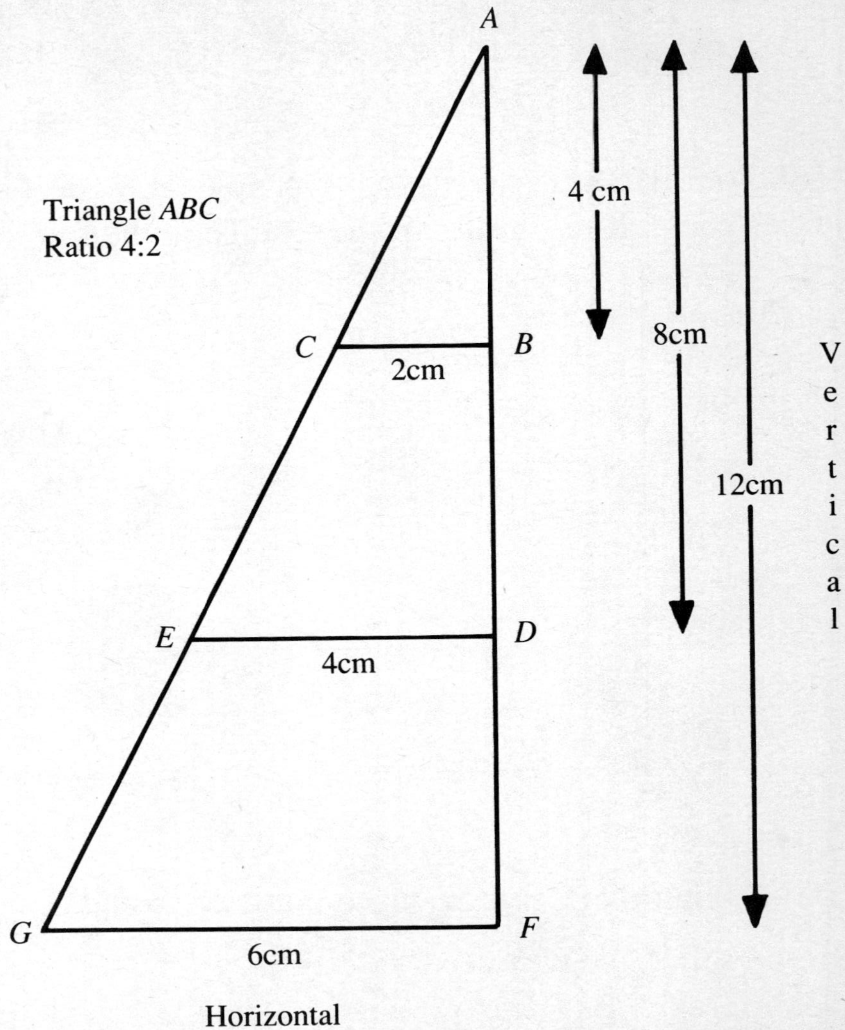

With your parent's help, write a rule that describes what you just learned.

The Odds
The Odds

Ask your parent to take turns with you rolling a number cube 15 times each. Keep a tally of the results on the frequency table below.

You need: a number cube and a pencil

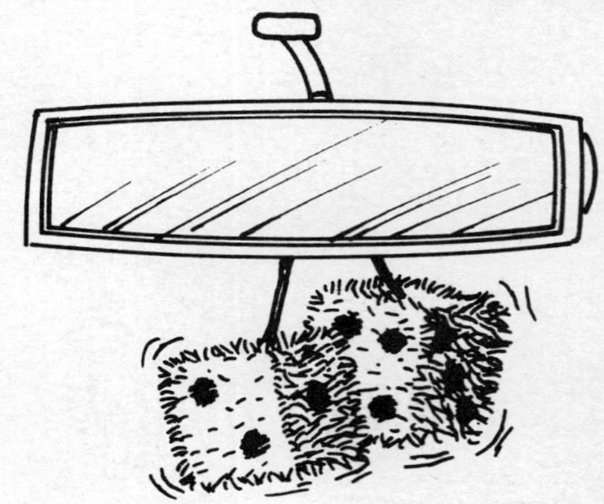

Number on cube	*Tally marks (𝍸 = 5)*	*Frequency*
1		
2		
3		
4		
5		
6		

What did you expect the results to be?

Did any number come up more often than others?

If you rolled the number cube another 30 times, would you expect the results to be different? Explain your answer. Now, try it.

Chances of a SIX

You need: a number cube, a pencil

If you rolled a number cube 30 times, estimate how many times you would roll number 6.

Ask your parent to estimate too.

Now calculate the probability of 6 appearing on a number cube in 30 rolls.

Number of faces on the number cube which show a 6 ↓

[ ] / [ ] × [ ] = [ ] to 1 (or) in 30

↑ Number of faces on the number cube

Number of rolls ↓

Probability of 6 appearing on a number cube in 30 rolls ↓

Roll a number cube 30 times. Ask your parent to keep a record of the numbers you roll.

How many times did you roll a 6? Considering the probability you previously calculated, were you lucky or unlucky with your sixes?

How could you make your results come closer to the calculated probability?

B or not B?

Write a different letter of the alphabet on 26 of the cards (you will have 6 cards left unused).

Put all 26 cards into a container and shake them.

Without looking at the cards, select one.

Is it *B*?

What is the probability of the letter being *B*?

To find the answer, enter the appropriate number in the boxes.

$$\frac{\text{Number of } B\text{s in the alphabet}}{\text{Number of letters in the alphabet}} = \frac{\square}{\square}$$

Return your card to the container and shake again.

Ask your parent to choose a letter. Is it *B*?

Explain to your parent the probability of a selection of *B* being made.

Leave *B* out of the container so that you have only 25 letters.

Pick a letter. Is it a vowel (*a, e, i, o, u*)?

Calculate the probability of the letter being a vowel.

You need: 4 index cards (or pieces of paper)—each card cut into eighths, a marker, container